AF411464

Hydrometeorological Extremes and Its Local Impacts on Human-Environmental Systems

Hydrometeorological Extremes and Its Local Impacts on Human-Environmental Systems

Editors

Jong-Suk Kim
Nirajan Dhakal
Changhyun Jun
Taesam Lee

MDPI • Basel • Beijing • Wuhan • Barcelona • Belgrade • Manchester • Tokyo • Cluj • Tianjin

Editors

Jong-Suk Kim
Department of Hydrology and
Water Resources
Wuhan University
Wuhan
China

Nirajan Dhakal
Environmental and Health
Sciences
Spelman College
Atlanta
United States

Changhyun Jun
School of Civil and
Environmental Engineering,
Urban Design and Studies
Chung-Ang University
Seoul
Korea, South

Taesam Lee
Dept. of Civil Eng.
Geyongsang National University
Jinju
Korea, South

Editorial Office
MDPI
St. Alban-Anlage 66
4052 Basel, Switzerland

This is a reprint of articles from the Special Issue published online in the open access journal *Atmosphere* (ISSN 2073-4433) (available at: www.mdpi.com/journal/atmosphere/special_issues/ hydrometeorological_extremes_impacts).

For citation purposes, cite each article independently as indicated on the article page online and as indicated below:

LastName, A.A.; LastName, B.B.; LastName, C.C. Article Title. *Journal Name* **Year**, *Volume Number*, Page Range.

ISBN 978-3-0365-3700-9 (Hbk)
ISBN 978-3-0365-3699-6 (PDF)

Contents

About the Editors

Jong-Suk Kim

Jong-Suk Kim is a Full Professor at the Department of Hydrology and Water Resources, Wuhan University, China. His research focuses on diagnostic analysis and assessment methodologies to understand hydroclimatic change, nonstationarity in hydroclimatic extremes, and an integrated view of climate and hydrologic variability and changes.

Nirajan Dhakal

Dr. Nirajan Dhakal achieved his Ph.D. at Auburn University, United States. Dr. Dhakal is an Associate Professor at the Environmental and Health Sciences, Spelman College. His research primarily focuses on surface hydrology, hydrological modeling, environmental impact assessments, and climate change.

Changhyun Jun

Dr. Changhyun Jun is an Assistant Professor at the School of Civil and Environmental Engineering, Chung-Ang University, Korea. Dr. Jun specializes in the areas of hydrological modeling, climate change, and disaster and risk management.

Taesam Lee

Dr. Taesam Lee is a Full Professor at the Department of Civil Engineering, Gyeongsang National University, Korea. Prof. Lee specializes in the areas of stochastic modeling, climate change, remote sensing, and water resource management.

Preface to "Hydrometeorological Extremes and Its Local Impacts on Human-Environmental Systems"

Extreme events of tropical typhoons in summer can cause many casualties, as well as tremendous amounts of social and financial loss. Such climate changes are expected to continue throughout the 21st century, and the intensity and frequency of typhoons over the Pacific Northwest region will increase. As a result, serious damage over East Asia is expected; thus, quantitative evaluations of the possible influences and establishment of a disaster-preventive system are urgent. Extreme hydrometeorological events are critically important not only for their episodic impacts, such as floods or droughts, but also for their significant contribution to seasonal freshwater supplies that maintain the integrity of human and natural systems. This Special Issue of *Atmosphere* focuses on hydrometeorological extremes and their local impacts on human–environment systems. Particularly, we accepted submissions on the topics of observational and model-based studies that could provide useful information for infrastructure design, decision making, and policy making to achieve our goals of enhancing the resilience of human–environment systems to climate change and increased variability.

Jong-Suk Kim, Nirajan Dhakal, Changhyun Jun, and Taesam Lee
Editors

Article

Copula-Based Drought Monitoring and Assessment According to Zonal and Meridional Temperature Gradients

Abudureymjang Otkur [1], Dian Wu [1], Yin Zheng [1], Jong-Suk Kim [1,*] and Joo-Heon Lee [2,*]

[1] State Key Laboratory of Water Resources and Hydropower Engineering Science, Wuhan University, Wuhan 430072, China; 2017301580078@whu.edu.cn (A.O.); 2020202060030@whu.edu.cn (D.W.); 2017301580034@whu.edu.cn (Y.Z.)

[2] Department of Civil Engineering, Joongbu University, Goyang-si 10279, Korea

* Correspondence: jongsuk@whu.edu.cn (J.-S.K.); leejh@joongbu.ac.kr (J.-H.L.)

Abstract: Drought is one of the most severe natural disasters. However, many of its characteristic variables have complex nonlinear relationships. Therefore, it is difficult to construct effective drought assessment models. In this study, we analyzed regional drought characteristics in China to identify their relationship with changes in meridional and zonal temperature gradients. Drought duration and severity were extracted according to standardized precipitation evapotranspiration index (SPEI) drought grades. Trends in drought duration and severity were detected by the Mann-Kendall test for the period of 1979–2019; they showed that both parameters had been steadily increasing during that time. Nevertheless, the increasing trend in drought severity was particularly significant for northwest and southwest China. A composite analysis confirmed the relationships between drought characteristics and temperature gradients. The northwest areas were relatively less affected by temperature gradients, as they are landlocked, remote from the ocean, and only slightly influenced by the land–ocean thermal contrast (LOC) and the meridional temperature gradient (MTG). The impacts of LOC and MTG on drought duration and severity were positive in the southwest region of China but negative in the northeast. As there was a strong correlation between drought duration and severity, we constructed a 2D copula function model of these parameters. The Gaussian, HuslerReiss, and Frank copula functions were the most appropriate distributions for the northeast, northwest, and southwest regions, respectively. As drought processes are highly complex, the present study explored the internal connections between drought duration and severity and their responses to meteorological conditions. In this manner, an accurate method of predicting future drought events was developed.

Keywords: copula function; drought duration; drought severity; land-ocean temperature contrast/meridional temperature gradient; standardized precipitation evapotranspiration index; China

Citation: Otkur, A.; Wu, D.; Zheng, Y.; Kim, J.-S.; Lee, J.-H. Copula-Based Drought Monitoring and Assessment According to Zonal and Meridional Temperature Gradients. *Atmosphere* **2021**, *12*, 1066. https://doi.org/10.3390/atmos12081066

Academic Editor: Hossein Tabari

Received: 19 July 2021
Accepted: 18 August 2021
Published: 20 August 2021

Publisher's Note: MDPI stays neutral with regard to jurisdictional claims in published maps and institutional affiliations.

1. Introduction

Drought has devastating impacts and may be complex, variable, and highly destructive to human-environment systems. Drought occurs in nearly all global climate regions, but it differs in each region [1,2]. It has a significant effect on water resources [3], agricultural production [4], economic activity [5], and ecosystems [6,7]. There have been severe, large-scale droughts on all continents except Antarctica [8,9]. Over the past two centuries, at least 40 droughts of long duration occurred in western Canada [10]. Europe has been affected by numerous major drought events over the past 30 years, especially in 1976, 1989, 1991, and 2003 [11]. Severe, extreme droughts occurred in the late 1990s in many parts of China. Persistent, severe, multiyear droughts have frequently occurred in northern, northeastern, northwestern, and southwestern China [12]. Hence, drought events have attracted considerable research attention in recent decades [13,14]. The Fifth IPCC Assessment Report stated that the global surface temperature rose by 0.85 °C (0.65–1.06 °C) between 1880 and 2012 [15]. As global surface temperatures continue to rise, the risk of drought will increase as well [16–20]. Recent research has focused mainly on the causes, intensity, frequency, impact, and countermeasures of droughts. To predict recurrences, quantitative analyses of the drought index (severity and duration) have been conducted using probability models [21]. However, the accurate evaluation of drought onset, ending, and intensity remains a major issue.

The causes of drought are highly complex. Thus, it is difficult to establish and quantify drought event indicators [22]. Drought indices are quantified using precipitation, evapotranspiration, soil moisture, streamflow, and groundwater level [1]. Multiple time-scale drought indicators, such as the effective drought index (EDI) [23], the standardized precipitation index (SPI) [24], and the standardized precipitation evapotranspiration index (SPEI) [25], have been developed. Precipitation is an important factors in drought monitoring, and its distribution pattern has been altered by climate change. China is shifting towards decreasing precipitation [26]. Therefore, drought may be exacerbated in China. For these reasons, it is necessary to elucidate the dynamic changes that occur in the climate system. Several studies have implemented large-scale temperature indices to explain the climate variability resulting from the changes in radiative force associated with increasing greenhouse gas (GHG) levels [27]. The equator-to-pole meridional temperature gradient (MTG) and the land-ocean temperature contrast (LOC) could help delineate changes in temperature gradients. The former is related to the Hadley cell and the westerlies [28], while the latter is associated with eddy strength and quasi-geostrophic flow at mid/high latitudes [29]. In recent years, several researchers have indicated that relative differences between sea and land temperature contribute to atmospheric circulation and climate anomalies [30].

Unlike the effects of other natural disasters such as earthquakes, floods, and winds, the impact of drought is usually significant but also difficult to characterize; it may have varying degrees of intensity. Hence, it is difficult to identify the start and end points of drought events [31]. To date, there is no comprehensive academic definition for drought [32]. Nevertheless, drought severity and duration have attracted widespread attention. They are considered the main parameters for estimating other drought characteristics such as intensity, and are critical metrics for real-time, long-term drought management [33,34]. Accurate understanding and analysis of drought severity and duration are essential for drought monitoring and risk/hazard assessment. Prior research was simplified by assuming that drought severity and duration were random, independent variables. However, Vicente-Serrano et al. [25] empirically demonstrated that this assumption is unreasonable. In fact, several characteristic drought variables are not independent and form complex nonlinear relationships.

As there are mutual dependencies among various drought parameters, separate analyses of drought duration and severity cannot reveal the significant correlations between them. Rather than using traditional univariate frequency analysis in drought assessment, a more effective approach towards describing drought characteristics is to determine the joint distribution of drought variables. Shiau and Shen [35], Bonaccorso [36], González and Valdés [37], Salas [38], and Cancelliere and Salas [39] proposed different methods to investigate the joint distribution of drought duration and severity/intensity. The multivariate distribution, known as the copula function, connects multiple variables conforming to any marginal distribution, obtains a joint distribution function, and describes the correlations among the variables. In this manner, the limitations of the aforementioned multivariate model are overcome [40]. The copula function is versatile and has been widely used in hydrological drought analysis [41]. Shiau [42] first applied the copula function to the analysis of meteorological drought events.

China is severely affected by drought. Disaster mitigation after drought events requires manpower, materials, and financial resources, as well as systematic and coordinated management strategies. Under climate change, hydrometeorological phenomena such as drought are substantially altered. In the present study, we used the copula method to explore the correlations among internal drought characteristics, the relationship between the east–west and north–south temperature gradients in the northern hemisphere, and the occurrence of drought in China. Drought magnitude and occurrence may be predicted based on the correlations among drought characteristics and meteorological conditions. The following questions were addressed in this study: (1) What are the key drivers of long-term changes in the characteristics of drought in China? (2) What are the associations between the drought characteristics and the temperature gradients in China? and (3) What are the connections between the temperature gradient patterns and regional drought in China? The present study also proposed copula-based drought monitoring and assessment approaches that help clarify the potential impact of climate systems on the extreme drought events across China. Figure 1 shows the research framework for this study.

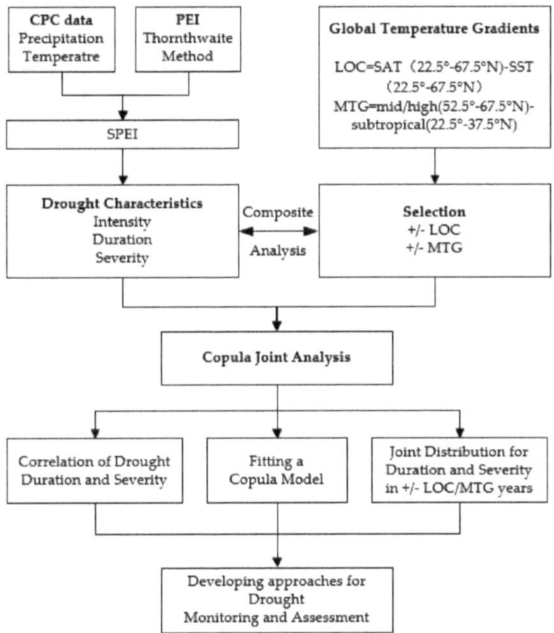

Figure 1. Study framework for copula-based drought monitoring and assessment using zonal and meridional circulations.

2. Materials and Methods

2.1. Study Area and Data

The datasets were provided by the National Oceanic and Atmospheric Administration (NOAA) (https://psl.noaa.gov/data/gridded/index.html, accessed on 17 August 2021). The monthly land surface temperatures were acquired from GHCN_CAMS Gridded 2 m Temperature (Land) dataset at $0.5° \times 0.5°$ resolution. The monthly sea surface temperatures were obtained from the NOAA Extended Reconstructed Sea Surface Temperature (SST) V5 dataset at $2° \times 2°$ resolution. Daily precipitation and temperature data comprised parts of the product suite generated by the climate prediction center (CPC) Unified Precipitation Project that is currently underway at NOAA CPC. All data used $0.5°$ latitude $\times$ $0.5°$ longitude as a unit. The datasets used in this study covered the period from 1979 to 2019. The composite meridional wind anomalies from the National Centers for Environmental Prediction/National Center for Atmospheric Research (NCEP/NCAR) re-analysis were provided by the NOAA.

The present study focused on the links between the east–west and north–south temperature gradients in the northern hemisphere and the regional droughts in China. As a result, the main area of research is China. Three regions in China were selected to determine the relationships between the drought characteristics and the global temperature gradients and included the northeastern (124°41′ E–130°44′ E, 44°43′ N–50°59′ N), northwestern (84°36′ E–90°44′ E, 40°57′ N–46°54′ N), and southwestern (100°08′ E–106°11′ E, 22°28′ N–28°30′ N) parts of China, respectively (Figure 2).

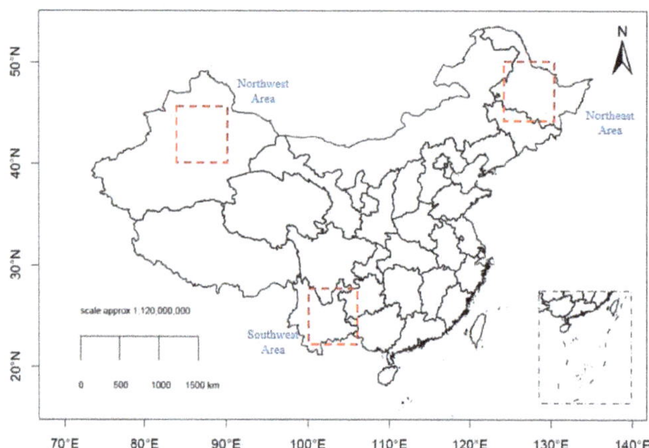

Figure 2. Study area comprising the northeastern, northwestern, and southwestern parts of China.

2.2. Standardized Precipitation Evapotranspiration Index

Drought indices such as the standardized precipitation index (SPI) and the standardized precipitation evapotranspiration index (SPEI) are commonly used to monitor meteorological drought [25,43]. SPI is used to monitor droughts based only on precipitation data, and does not consider the impact of other factors on evapotranspiration [43]. However, SPEI proposed by Vicente-Serrano evaluates drought based on both evapotranspiration and precipitation [25]. In this study, SPEI was applied to identify global temperature indices and drought characteristics in China. The SPEI drought ratings are shown in Table 1.

Table 1. Drought categories based on the SPEI index.

Category	SPEI Index
Normal	$(-0.5, +\infty)$
Slight drought	$(-1.0, -0.5)$
Moderate drought	$(-1.5, -1.0)$
Severe drought	$(-2.0, -1.5)$
Extreme drought	$(-\infty, -2.0]$

The SPEI calculation procedure is summarized as follows. The monthly potential evaporation (PET) is calculated according to the Thornthwait method, and the difference between monthly precipitation and potential evaporation was calculated as follows:

$$D_i = P_i - PET_i \tag{1}$$

where P_i is the precipitation and PET_i is the potential evaporation.

The cumulative water gain/loss sequence in climatological significance at various time scales was calculated as follows:

$$D_n^k = \sum_{i=0}^{k-1}(P_{n-i} - PET_{n-i}), \; n \geq k \tag{2}$$

where k is the monthly time scale and n is the month.

The established data sequence was fitted with three parameters by the log-logistic probability density function as follows:

$$f(x) = \frac{\beta}{\alpha}(\frac{x-\gamma}{\alpha})^{\beta-1}[1+(\frac{x-\gamma}{\alpha})^{\beta}]^{-2} \tag{3}$$

where α is the scale parameter, β is the shape parameters, and γ is the origin parameters. These were obtained by L-moment parameter estimation. Hence, the cumulative probability for a given time scale can be calculated as follows:

$$F(x) = [1 + \frac{\alpha}{x-\gamma}]^{-1} \tag{4}$$

The results were processed by standardized normal distribution and the solution was obtained as follows:

$$SPEI = W - \frac{C_0 - C_1 W + C_2 W^2}{1 + d_1 W + d_2 W^2 + d_3 W^3} \tag{5}$$

$$W = \sqrt{-2\ln(P)}, \ P \leq 0.5 \tag{6}$$

where $P = 1 - F(x)$; when $P > 0.5$, P becomes $1 - P$. $C_0 = 2.515517$, $C_1 = 0.802853$, $C_2 = 0.010328$, $d_1 = 1.432788$, $d_2 = 0.189269$, and $d_3 = 0.001308$.

2.3. Definition of Drought Characteristics

The schematic diagram in Figure 3 defines drought. A meteorological drought event is generally defined as the period in which the meteorological variable of interest is below a certain cutoff value. This interception level may either be constant or variable. Therefore, the important indicators of drought events define drought duration and severity.

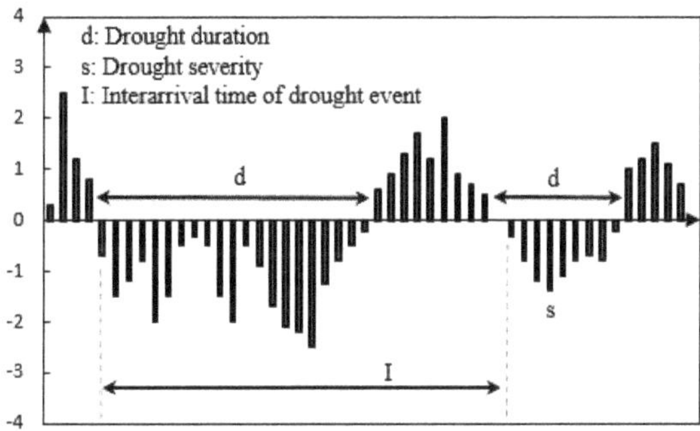

Figure 3. Definition of drought events used in the present study.

According to SPEI series, drought is identified by presuming that a drought period is continuous over time when SPEI < −0.5 [42]. Therefore, the period wherein SPEI < −0.5 is defined as the duration of moderate drought. The absolute value of the cumulative SPEI over the drought duration indicates the drought severity.

2.4. Trend Analysis

A trend analysis such as a Mann-Kendall (MK) test is usually required for intuitive determination of the tendency in time series data. The nonparametric MK statistical method is recommended by the World Meteorological Organization (WMO), and is widely used to distinguish whether a natural process is in natural fluctuation or has a definite change trend [44]. However, the MK rank correlation test is more appropriate for non-normally distributed hydrometeorological data. No specific distribution test is required for the data series, extreme values can be tested in the trend test, the series can include missing values, and the relative order of magnitude may be analyzed rather than the number itself. In this way, values below the detection range and trace values may be included in the analysis. Moreover, linear trends need not be specified in time series analysis.

The MK test is often used to detect precipitation and drought frequency under the influence of climate change. The trend detection method was previously reported [40]. The alternative hypothesis H_1 is a bilateral test. For all the distributions of i, j $\leq$ n and i $\neq$ j and the distributions of X_i and X_j differ from each other. The MK correlation coefficient between two variables is known as the MK statistic S. It is calculated as follows:

$$S = \sum_{i=1}^{n-1} \sum_{j=i+1}^{n} sgn(X_j - X_i) \tag{7}$$

where sgn() is a symbolic function shown in Equation (8):

$$sgn(X_j - X_i) = \begin{cases} 1 \text{ if} (X_j - X_i) > 0 \\ 0 \text{ if} (X_j - X_i) = 0 \\ -1 \text{ if} (X_j - X_i) < 0 \end{cases} \tag{8}$$

where S is an approximately normal distribution with a mean = 0 and variance calculated as shown in Equation (9) below:

$$Var(S) = \frac{n(n-1)(2n-5) - \sum_{i=1}^{n} t_i(i-1)(2i+5)}{18} \tag{9}$$

where t is the range of any given node. When n > 10, Z_C converges to the standard normal distribution and is calculated as shown in Equation (10) below:

$$Z_c = \begin{cases} \frac{S-1}{\sqrt{Var(S)}} & S > 0 \\ 0 & S = 0 \\ \frac{S+1}{\sqrt{Var(S)}} & S < 0 \end{cases} \tag{10}$$

In the foregoing calculation, the original hypothesis H_0 will be rejected when $|Z_C| \geq Z_{1-\alpha/2}$. In this case, there is a significant upward or downward trend in the current time series data α confidence level. If $Z_C > 0$, it shows an upward trend. If $Z_C < 0$, it shows a downward trend. $\pm Z_{1-\alpha/2}$ is the $(1 - \alpha/2)$ quantile of the standard normal distribution, and α is the test confidence level of the test. When $|Z_C| \geq 1.64$, 1.96, and 2.58, the α confidence values are 0.1, 0.05, and 0.01, respectively. Here, the MK test was used to analyze the drought characteristic across China, and the tendency was determined from the results of the p and z values. The trend is not significant when $p > 0.05$. However, the difference between the result and the zero hypothesis is evident when $p < 0.05$. Thus, the p-value can indicate whether the SPEI of the area has a significant change trend. The direction of trend is then determined based on the positive and negative S values.

2.5. Meridional and Zonal Temperature Gradients

The temperature difference between land and ocean is an important characterization of the land-ocean thermal contrast that significantly influences the regional climate. China is located on the Eurasian continent and faces the Pacific Ocean. Hence, the difference between the land and the ocean temperature has an enormous impact on the climate. Temperature differences may be expressed by the LOC index which is defined as the longitudinal variation in the difference between the average land surface air temperature (SAT) and the average sea surface temperature (SST). It represents the difference in temperature between the land surface and the ocean attributed to global warming [30]. The LOC index may be used to detect changes in seasonal precipitation caused by water vapor movement.

Equation (11) is based on the basic principle of the LOC index, calculated in the present study:

$$\text{LOC} = \text{SAT}(22.5°\text{–}67.5°\ \text{N}) - \text{SST}(22.5°\text{–}67.5°\ \text{N}) \tag{11}$$

The meridional temperature gradient is caused by differences in atmospheric pressure between high and low latitudes. North–south atmospheric circulation redistributes water vapor and energy across various regions. The observed hydroclimate also varies spatiotemporally. A large proportion of the global population lives in the Asian monsoon region. Consequently, it is crucial to understand the movement of the intertropical convergence zone (ITCZ) and the north subtropical high to assess their impact on water vapor transport, energy transfer, and interregional global climate [41]. In academia, however, there is a lack of understanding of the history and laws of the ITCZ. Furthermore, the north subtropical highs and their locations have not been precisely defined. Based on the extent of the temperate zone near the tropics, the subtropical zone was defined here as 22.5–37.5° N. Equation (12) is the formula based on the basic principle of the MTG index calculated in the present study:

$$\text{MTG} = \text{mid/high}((52.5°\text{–}67.5°\ \text{N}) - \text{subtropical}(22.5°\text{–}37.5°\ \text{N}) \tag{12}$$

2.6. Copula Functions

Copula functions [45] link univariate distributions to form multivariate distributions. They separate the dependence effects from the marginal distributions and facilitate model construction and study [42]. Copula functions also maintain the uniqueness of the probability distribution function of a single variable and retain valuable information when the model is transformed [46]. They have been recently applied to complex hydrologic phenomena such as drought.

Genest and Mackay proposed the Archimedean copula function in 1986 [47]. It is expressed as follows:

$$C(\mu_1,\dots,\mu_n,\dots,\mu_N) = \varphi^{-1}(\varphi(\mu_1),\dots,\varphi(\mu_n),\dots,\varphi(\mu_N)) \tag{13}$$

where φ is the generating function that fully determines the Archimedean copula function. The Archimedean copula function family includes mainly symmetrical and asymmetrical copula functions. In this study, two symmetrical functions named G-H and Frank were used. Their expressions are shown below:

$$C(u,v) = \exp\left\{-[-\ln u]^{\frac{1}{\theta}}\right\},\ \theta \in [1,+\infty) \tag{14}$$

$$C(u,v) = -\frac{1}{\theta}\ln\left[1+\frac{(e^{-\theta v}-1)(e^{-\theta u}-1)}{e^{-\theta}-1}\right],\ \theta \neq 0 \tag{15}$$

Ellipse copula functions are derived from the ellipse distribution. The elliptic copula function family consists mainly of the Gaussian copula function and t-copula functions. The Gaussian copula function is derived from the multivariate normal distribution as shown below:

$$C(u_1,\dots,u_n,\dots,u_{N,\varrho}) = \varphi_\rho(\varphi^{-1}(u_1),\dots,\varphi^{-1}(u_n),\dots,\varphi^{-1}(u_N)) \tag{16}$$

$$c(u_1,\dots,u_n,\dots,u_{N,\varrho}) = |\varrho|^{-\frac{1}{2}}\exp\left\{\frac{-1}{2}\varsigma^{\mathrm{T}}(\varrho^{-1}-I)\varsigma\right\} \tag{17}$$

where φ^{-1} is the inverse function of the standard normal distribution function, φ_ϱ is the standard multivariate normal distribution, ϱ is the correlation coefficient matrix between variables and is a symmetric positive definite matrix. I is the identity matrix, and $\varsigma n = \varphi^{-1}(u_n)$.

The t-copula or Student's copula is derived from the t distribution as shown below:

$$C(u_1,\cdots,u_n,\cdots,u_N,\varrho) = T_{\varrho,v}(t_v^{-1}(u_1),\cdots,t_v^{-1}(u_n),\cdots,t_v^{-1}(u_N)) \tag{18}$$

$$c(u_1,\cdots,u_n,\cdots,u_N,\varrho,v) = |\varrho|^{\frac{1}{2}}\frac{\Gamma(\frac{N+v}{2})[\Gamma(\frac{v}{2})]^N(1+\frac{1}{v}\varsigma^{-1}\varrho^{-1}\varsigma)^{\frac{-v+N}{2}}}{[\Gamma(\frac{N+v}{2})]^N\Gamma(\frac{v}{2})\prod_{n=1}^{N}(1+\frac{\varsigma_n^2}{v})^{\frac{-v+2}{2}}} \tag{19}$$

where the $T_{\varrho,v}$ is the t standard multivariate distribution, v is the degree of freedom, ϱ is the correlation coefficient matrix between variables and is a symmetric positive definite matrix, $|\varrho|$ is the absolute value of the determinant of ϱ, t_v^{-1} is the inverse function of the one-variable t distribution function with degrees of freedom, and $\varsigma_n = \varphi^{-1}(u_n)$.

Extreme value copulas mainly include the Galambos and HuslerReiss functions. Their formulae are displayed below:

$$C(u,v) = uv\cdot\exp\{[(-\ln u)^{-\theta}+(-\ln v)^{-\theta}]^{-\frac{1}{\theta}}\},\ \theta > 0 \tag{20}$$

$$C(u,v) = \exp\{-\tilde{u}\Phi[\frac{1}{\alpha}+\frac{1}{2}\alpha\ln(\frac{\tilde{v}}{\tilde{u}})]-\tilde{v}\Phi[\frac{1}{\alpha}+\frac{1}{2}\alpha\ln(\frac{\tilde{v}}{\tilde{u}})]\} \tag{21}$$

where $\alpha \geq 0$, $\tilde{u}= -\ln u$, $\tilde{v}= -\ln v$, and Φ is the standard normal distribution function.

3. Results

3.1. Changes in Large-Scale Temperature

In this study, the large-scale temperature indices LOC and MTG were used to analyze the dynamic changes within the climate system. Figure 4 represents the temporal trend in the two normalized temperature gradient indices in the springtime of the years 1979 to 2019.

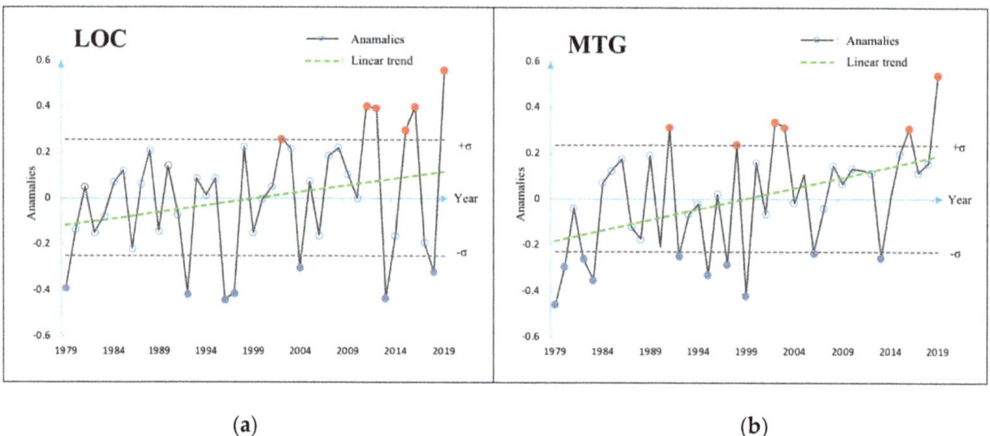

(a)   (b)

Figure 4. Time series of global temperature gradients. (**a**) LOC index. (**b**) MTG index during the period from 1979 to 2019. Green dotted line represents the linear trend of the indices. Gray dashed line indicates the +1 σ and −1 σ. For each gradient, red dots indicate significant positive events while blue dots indicate significant negative events.

When LOC > 0, the land temperature is higher than the sea temperature. When LOC < 0, the opposite is true. When LOC approaches zero, there is minimal difference between land and sea temperature. The MTG index reflects the temperature difference between high and low latitudes. Current research showed that the Arctic region is relatively more sensitive to GHG-induced global warming than previously expected [48]. The decrease in the temperature difference between high and low latitudes was apparent from the decline in the MTG index. In the present study, points higher than the quantile line were selected as positive years while those lower than the quantile line were selected as negative. Hence, the positive LOC years were 2002, 2011, 2012, 2015, 2016, and 2019 while the negative LOC years were 1979, 1992, 1996, 1997, 2004, 2013, and 2018. Similarly, 1991, 1998, 2002, 2003, 2016, and 2019 were selected as positive MTG years while 1979, 1980, 1982 ,1983, 1992, 1995, 1997, 1999, 2006 and 2013 were selected as negative MTG years.

3.2. Composite Anomalies of Drought Characteristics Associated with LOC/MTG

To quantify the influences of the temperature gradients on the study area, the SPEI index was calculated and the drought duration and severity were extracted based on the drought definition. The average duration and severity observed in the period of 1979–2019, and the negative and positive LOC/MTG years in the southwestern, northwestern, and northeastern regions of China, were enumerated. We obtained the results of the composite analysis of the drought characteristics for the positive and negative LOC/MTG years. The variations in the drought characteristics of each region were associated with four indices and are displayed in Figure 5. In northeastern and southwestern China, the influences of positive/negative LOC and MTG on drought duration and severity were consistent. Compared with the average years, the positive LOC, negative LOC, and positive MTG years had significantly weaker effects on drought duration and severity in the northeast but significantly stronger effects on drought duration and severity in the southwest. In northwestern China, positive LOC increased drought duration and decreased drought severity. By contrast, negative LOC and positive MTG decreased drought duration and increased drought severity in northwestern China. In the negative MTG years, drought severity in the northwest had a maximum relative increase of 15.7%. Positive land-ocean temperature difference had the most significant impact on drought duration in the southwest. There, the average change in annual mean duration during the positive LOC years within the period of 1979–2019 increased by 24.3%. The northeast had the most significant trend in positive MTG indicators with a decrease of 16.1%. The changes in LOC and MTG in the northwest were relatively small.

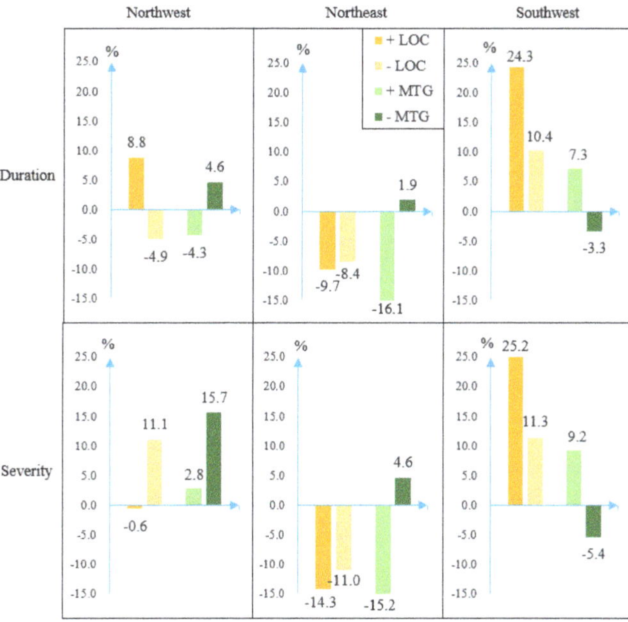

Figure 5. Composite analysis of drought characteristics with temperature gradients.

The observed changes in drought severity in the northeast and southwest were consistent with the drought duration for the LOC and MTG years. However, the southwest was significantly affected by positive LOC and increased by 25.2%. In general, drought severity was more strongly affected by LOC and MTG than drought duration.

3.3. Changes in Drought Duration and Severity in China

China is severely affected by frequent drought disasters [49,50]. Since the 1990s, crop damage caused by flood and drought disasters has significantly increased. The current area affected by drought and flooding is 1.4 times the multiyear average between 1950 and 2010 [51]. Hence, drought disasters pose serious threats to food and ecological security and have become one of the critical factors limiting the sustainable development of the society and the economy of China. Several severe drought events occurred in southern and central China in 2001, in southeastern and western China in 2007, and in southwest and northern China in 2011. The main reason for these droughts was the weakening of the favorable east–west zonal wind. Consequently, insufficient water vapor is brought from the ocean. Further, the north–south meridional wind, which brings dry inland air, may also be strengthened.

In the present study, we used daily precipitation and maximum data to calculate the SPEI index over a 12-month period and extracted the drought duration and severity characteristics. We applied the M-K test to elucidate the drought duration and severity trends. The trends of drought duration and severity in China between 1979 and 2019 are displayed in Figure 6. Figure 6a shows that except for an increasing trend in drought duration in northwestern China, there were no perceptible trends in any other areas. The rising trend in drought severity in northwestern China is also depicted in Figure 6b. Figure 6 shows a correlation between drought severity and drought duration. As a rule, drought severity increases with drought duration. Compared with drought duration, drought severity has more significantly changed over the past few decades.

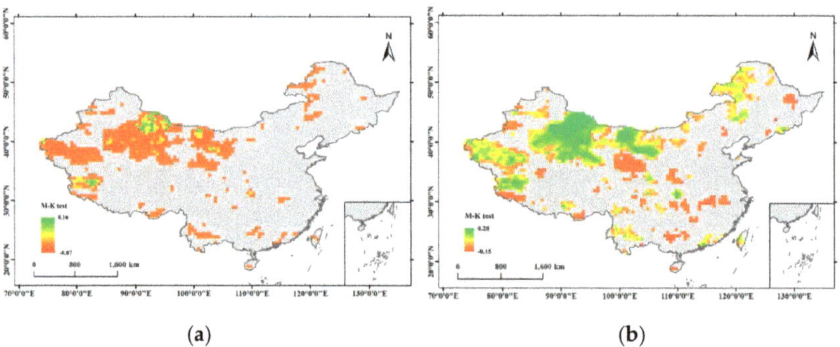

(a)   (b)

Figure 6. Long-term trend in drought components in China between 1979 and 2019. (**a**) Drought duration. (**b**) Drought severity.

3.4. Joint Distribution of Drought Duration and Severity

The characteristic drought variables such as duration and severity are closely related. Drought duration and severity are usually positively correlated. Pearson's correlation, Spearman's correlation, and Kendall coefficients measure the association between drought duration and severity. The strongest relationship between drought duration and severity was identified by Pearson's correlation coefficient. Figure 7 shows the relationship between drought duration and severity in the study area between 1979 and 2019.

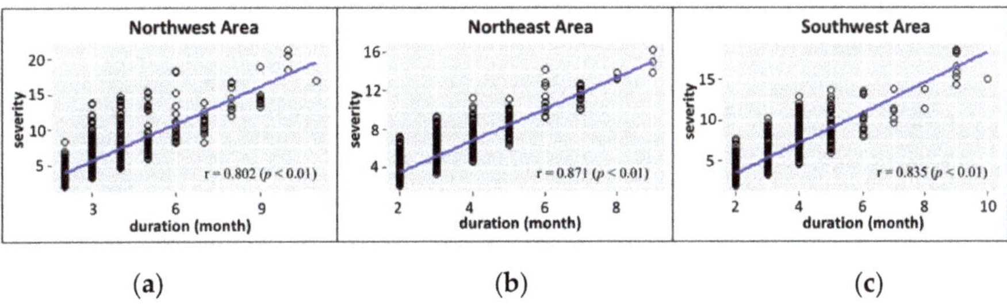

(**a**)   (**b**)   (**c**)

Figure 7. Relationship between drought duration and severity in the (**a**) southwest, (**b**) northeast, and (**c**) southwest.

The joint distribution of the drought characteristics was not clear. According to the copula function theory, we selected six copula functions to fit the joint distribution of drought duration and severity: Gaussian Copula and Student's copula in the elliptic copula function family, GH Copula and Frank copula in the Archimedes copula function family, and Galambos copula and HuslerReiss copula in the extreme copula function family (Table 2). The best-fitting copula model was selected to represent the joint probability model of drought duration and severity.

Table 2. Copula model-fitting table for the study area.

Area	Category of Copula Function	Parameter Estimates	Goodness of Fit Evaluation Standard	
			AIC	BIC
Northwest	Gaussian copula	0.8271	−601.2871	−596.9415
	Student's copula	0.8271	−599.1858	−590.4946
	GH copula	2.345	−546.6435	−542.2979
	Frank copula	8.183	−573.2539	−568.9083
	Galambos copula	1.645	−554.9031	−550.5575
	HuslerReiss copula	2.252	−565.7303	−561.3846
Northeast	Gaussian copula	0.8319	−820.555	−815.905
	Student's copula	0.8337	−822.6324	−813.3318
	GH copula	2.701	−907.3378	−902.6875
	Frank copula	8.604	−791.2964	−786.6461
	Galambos copula	2.003	−909.7965	−905.1462
	HuslerReiss copula	2.635	−910.1736	−905.5233
Southwest	Gaussian copula	0.8291	−775.5213	−773.5139
	Student's copula	0.8291	−773.5139	−764.3061
	GH copula	2.436	−737.7691	−733.1652
	Frank copula	8.816	−784.6019	−779.998
	Galambos copula	1.727	−744.8849	−740.2809
	HuslerReiss copula	2.335	−757.1334	−752.5294

Based on the AIC and BIC values in the table, the Frank copula function model fit best for the southwest. The HuslerReiss copula model was best suited for northeast, while the Gaussian copula was most appropriate for the northwest. Therefore, the Gaussian copula, HuslerReiss copula, and Frank copula functions were selected for the 2D copula joint distribution model of the drought characteristics in the three foregoing regions.

Figure 8 shows contour maps of the 2D joint distributions of drought duration and severity and demonstrates the strong dependence between the two drought characteristics. For example, the figure shows joint distributions for various combinations of drought duration and severity in positive (negative) LOC years and positive (negative) MTG years for northwestern, northeastern, and southwestern China.

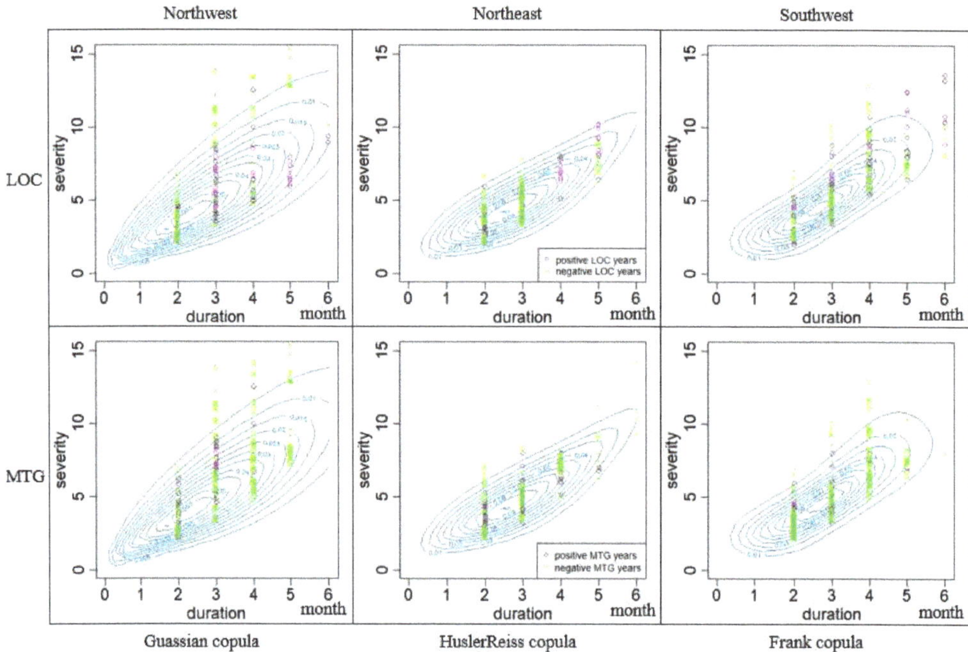

Figure 8. Copula distributions of drought duration and severity in positive (negative) LOC/MTG years.

4. Discussion and Conclusions

In this study, SPEI was used as a meteorological drought indicator. Drought duration and intensity served to define drought events based on daily CPC temperature and precipitation data between 1979 and 2019. The relationships between zonal and meridional temperature gradients (LOC/MTG) and drought characteristics (duration and severity) were also investigated. Hence, we aimed to forecast the occurrence and magnitude of drought events based upon drought characteristic correlations and their responses to meteorological conditions.

Trend analyses of drought duration and severity show that their long-term changes were positively correlated and increasing. Drought severity has changed more significantly than drought duration over the past few decades. In China, the observed increases in drought were localized in the northwestern and southwestern regions. These findings were consistent with those reported by Zeng et al. [52]. Drought events pose substantial threats to the environment and human society in those areas. As drought duration and severity have steadily increased, China may experience even more severe drought events going forward.

A composite analysis of drought characteristics temperature gradients revealed that drought severity is relatively more affected by LOC and MTG than drought duration. In northwestern China, the changes in LOC and MTG were comparatively minor. The northwestern region of China is far inland and remote from the ocean. Hence, the influences of land-ocean thermal contrast and the meridional temperature gradient are relatively small there. Nevertheless, drought severity has increased by 15.7% in this region because of negative MTG. Compared with other temperature gradients, the northwestern region is more strongly affected by negative LOC and MTG. The northeastern and southwestern regions of China are relatively closer to the ocean and are, therefore, more strongly influenced by land-ocean thermal contrast. Whereas LOC and MTG reduce drought duration and severity in the northeast, LOC and MTG enhance drought duration and severity in the southwest. This discrepancy could be explained by the relative differences in the geography of these two areas as well as other factors that remain to be determined. Positive MTG has the most negative effect on the northeastern region while positive LOC has the most positive effects on the southwestern region.

Univariate analyses do not objectively or comprehensively describe drought events. Therefore, multivariate analyses are applied for this purpose. However, traditional multivariate analyses are used mainly in scenarios where each variable has a normal marginal distribution. The copula function is recommended for the construction of a joint marginal distribution because it can effectively describe the correlations among variables and has considerable advantages over traditional multivariate analyses. The copula function plots a 2D copula distribution with drought duration and severity. The present study compared AIC values for six copula functions. It was established that the Gaussian, HuslerReiss, and Frank copula functions were suitable for the northwestern, northeastern, and southwestern areas of China, respectively. The marginal distribution function of drought characteristics facilitates calculation of the conditional probabilities of drought duration and severity in LOC/MTG years.

The foregoing analyses showed that the northwestern and southwestern regions of China may encounter more severe and frequent drought events in the future. However, the increases in drought severity will be more significant than the increases in drought duration. The copula function methods revealed strong positive correlations between drought duration and severity. As drought severity and duration intensify, China will be challenged by even more serious drought disasters that could be devastating to agriculture, the economy, and human society. Moreover, worsening climate warming may aggravate these changes. Composite analyses of drought characteristics and temperature gradients disclose the responses of the various regions of China to changing meteorological conditions. As drought processes are highly complex, their short-term impact may not be immediately apparent. For this reason, exploring the connections among drought variables and their responses to climate change might enable accurate prediction of the occurrences and magnitudes of drought events in the future.

Author Contributions: Conceptualization, Resources, Methodology, Formal analysis, Writing—original draft, A.O., D.W., Y.Z. and J.-S.K.; Conceptualization, Writing—review & editing, J.-H.L. All authors have read and agreed to the published version of the manuscript.

Funding: We appreciate the support of the State Key Laboratory of Water Resources and Hydropower Engineering Science, Wuhan University. This work was also supported by the National Research Foundation of Korea (NRF) grant funded by the Korea government (MSIT) (No.2021R1A2C1013190).

Institutional Review Board Statement: Not applicable.

Informed Consent Statement: Not applicable.

Data Availability Statement: Data set available on request to corresponding authors.

Conflicts of Interest: The authors declare no conflict of interest.

References

1. Mishra, A.K.; Singh, V.P. A review of drought concepts. *J. Hydrol.* **2010**, *391*, 202–216. [CrossRef]
2. Tallaksen, L.M.; Van Lanen, H.A.J. *Hydrological Drought-Processes and Estimation Methods for Streamflow and Groundwater;* Developments in Water Science; Elsevier Science BV: Amsterdam, The Netherlands, 2004.
3. Arash, M.R.; Bijan, G.; Davar, K.; Zahra, G.; Samira, A.A. Integrated meteorological and hydrological drought model: A management tool for proactive water resources planning of semi-arid regions. *Adv. Water Resour.* **2017**, *107*, 336–353.
4. Luo, D.; Ye, L.L.; Sun, D.C. Risk evaluation of agricultural drought disaster using a grey cloud clustering model in Henan province, China. *Int. J. Disaster Risk Reduct.* **2020**, *49*, 101759. [CrossRef]
5. Muneta, Y.; Hiroaki, I.; Sawada, Y.; Suzuki, Y.S.; Toshio, K.; Asif, N.; Cheema, M.J.M. A multi-sector multi-region economic growth model of drought and the value of water: A case study in Pakistan. *Int. J. Disaster Risk Reduct.* **2020**, *43*, 101368.
6. Crausbay, S.; Ramirez, A.R.; Carter, S.L.; Cross, M.S.; Hall, K.R.; Bathke, D.J.; Betancourt, J.L.; Colt, S.; Cravens, A.E.; Dalton, M.S.; et al. Defining Ecological Drought for the Twenty-First Century. *Bull. Am. Meteorol. Soc.* **2017**, *98*, 2543–2550. [CrossRef]
7. Bradford, J.B.; Schlaepfer, D.R.; Lauenroth, W.K.; Palmquist, K.A. Robust ecological drought projections for drylands in the 21st century. *Glob. Chang. Biol.* **2020**, *26*, 3906–3919. [CrossRef] [PubMed]

8. Le Comte, D. Weather highlights around the world. *Weatherwise* **1995**, *48*, 20–22. [CrossRef]
9. Le Comte, D. Weather highlights around the world. *Weatherwise* **1994**, *47*, 23–26. [CrossRef]
10. Phillips, D. *The Climates of Canada*; Environment Canada: Ottawa, ON, 1990; p. 176.
11. Feyen, L.; Dankers, R. Impact of global warming on streamflow drought in Europe. *J. Geophys. Res. Atmos.* **2009**, *114*, D17. [CrossRef]
12. Ayantobo, O.O.; Li, Y.; Song, S.; Yao, N. Spatial comparability of drought characteristics and related return periods in mainland China over 1961–2013. *J. Hydrol.* **2017**, *550*, 549–567. [CrossRef]
13. Ding, Y.; Hayes, M.J.; Widhalm, M. Measuring economic impacts of drought: A review and discussion. *Disaster Prev. Manag.* **2011**, *20*, 434–446. [CrossRef]
14. Li, X.Y.; Waddington, S.R.; Dixon, J.; Joshi, A.K.; De Vicente, M.C. The relative importance of drought and other water-related constraints for major food crops in South Asian farming systems. *Food Secur.* **2011**, *3*, 19–33. [CrossRef]
15. Seneviratne, S. Changes in climate extremes and their impacts on the natural physical environment: An overview of the IPCC SREX report. In *Managing the Risks of Extreme Events and Disasters to Advance Climate Change Adaptation*; Cambridge University Press: Cambridge, UK; New York, NY, USA, 2012.
16. Jefferson, M. IPCC fifth assessment synthesis report: "Climate change 2014: Longer report". *Crit. Anal. Technol. Forecast. Soc. Change* **2014**, *92*, 362–363. [CrossRef]
17. Stewart, I.T.; Rogers, J.; Graham, A. Water security under severe drought and climate change: Disparate impacts of the recent severe drought on environmental flows and water supplies in Central California. *J. Hydrol. X* **2020**, *7*, 100054. [CrossRef]
18. Martínez-Valderrama, J.; Ibanez, J.; Ibanez, M.A.; Alcala, F.J.; Sanjuan, M.E.; Ruiz, A.; Del Barrio, G. Assessing the sensitivity of a Mediterranean commercial rangeland to droughts under climate change scenarios by means of a multidisciplinary integrated model. *Agric. Syst.* **2021**, *187*, 103021. [CrossRef]
19. Noorisameleh, Z.; Khaledi, S.; Khaledi, A.; Firouzabadi, P.Z.; Gough, W.A.; Mirza, M.M.Q. Comparative evaluation of impacts of climate change and droughts on river flow vulnerability in Iran. *Water Sci. Eng.* **2021**, *13*, 265–274. [CrossRef]
20. Rehana, J.; Naidu, G.S. Development of hydro-meteorological drought index under climate change—Semi-arid river basin of Peninsular India. *J. Hydrol.* **2021**, *594*, 125973. [CrossRef]
21. Dai, A.G. Increasing drought under global warming in observations and models. *Nat. Clim. Change* **2012**, *3*, 52–58. [CrossRef]
22. Dai, A.G. Drought under global warming: A review. *Wiley Interdiscip. Rev. Clim. Change* **2011**, *2*, 45–65. [CrossRef]
23. Byun, H.R.; Wilhite, D.A. Objective quantification of drought severity and duration. *Am. Meteorol. Soc. J. Clim.* **1999**, *12*, 2747–2756. [CrossRef]
24. McKee, T.B.; Doesken, N.J.; Kleist, J. The relationship of drought frequency and duration to time scales. In Proceedings of the Eighth Conference on Applied Climatology, Anaheimm, CA, USA, 17–23 January 1993.
25. Vicente-Serrano, S.M.; Beguería, S.; López-Moreno, J.I. A Multiscalar Drought Index Sensitive to Global Warming: The Standardized Precipitation Evapotranspiration Index. *J. Clim.* **2010**, *23*, 1696–1718. [CrossRef]
26. Dore, M.H.I. Climate change and changes in global precipitation patterns: What do we know? *Environ. Int.* **2005**, *31*, 1167–1181. [CrossRef] [PubMed]
27. Zou, X.; Zhai, P.; Zhang, Q. Variations in droughts over China: 1951–2003. *Geophys. Res. Lett.* **2005**, *32*, 353–368. [CrossRef]
28. Jeong, M.S.; Kim, J.-S.; Oh, S.M.; Moon, Y.-I. Sensitivity of summer precipitation over the Korean Peninsula to temperature gradients. *Int. J. Clim.* **2015**, *35*, 836–845. [CrossRef]
29. Jain, S.; Lall, U.; Mann, M.E. Seasonality and Interannual Variations of Northern Hemisphere Temperature: Equator-to-Pole Gradient and Ocean–Land Contrast. *J. Clim.* **1999**, *12*, 1086–1100. [CrossRef]
30. Xu, H. Arctic climate is more sensitive to global warming than expected. In *Resources, Environment and Engineering*; Xie, L.Q., Ed.; London, UK, 2014; Volume 28, pp. 232–233.
31. Quiring, S.M.; Papakryiakou, T.N. An evaluation of agricultural drought indices for the Canadian prairies. *Agric. For. Meteorol.* **2003**, *118*, 49–62. [CrossRef]
32. Funk, C.; Shukla, S. Drought early warning—definitions, challenges, and opportunities. In *Drought Early Warning and Forecasting*; Elsevier: Amsterdam, The Netherlands, 2020; pp. 23–42.
33. Kim, T.-W.; Valdes, J.B.; Yoo, C. Nonparametric approach for bivariate drought characterization using Palmer drought index. *J. Hydrol. Eng.* **2006**, *11*, 134–143. [CrossRef]
34. Hao, Z.; Singh, V.P. Entropy based method for bivariate drought analysis. *J. Hydrol. Eng.* **2013**, *18*, 780–786. [CrossRef]
35. Shiau, J.T.; Shen, H.W. Recurrence analysis of hydrologic droughts of differing severity. *J. Water Resour. Plan. Manag.* **2001**, *127*, 30–40. [CrossRef]
36. Bonaccorso, B.; Cancelliere, A.; Rossi, G. An analytical formulation of return period of drought severity. *Stoch. Environ. Res. Risk Assess.* **2003**, *17*, 157–174. [CrossRef]
37. González, J.; Valdés, J.B. Bivariate Drought Recurrence Analysis Using Tree Ring Reconstructions. *J. Hydrol. Eng.* **2015**, *8*, 247–258. [CrossRef]
38. Salas, J.D.; Fu, C.J.; Cancelliere, A.; Dustin, D.; Bode, D.; Pineda, A.; Vincent, E. Characterizing the Severity and Risk of Drought in the Poudre River, Colorado. *J. Water Resour. Plan. Manag.* **2005**, *131*, 383–393. [CrossRef]
39. Cancelliere, A.; Salas, J.D. Drought Probabilities and Return Period for Annual Streamflows Series. *J. Hydrol.* **2010**, *391*, 77–89. [CrossRef]

40. Yang, X.Q.; Xiao, J. Analysis of precipitation change trend and mutation in Hainan Island based on Mann Kendall. *Flood Control. Drought Relief* **2020**, *30*, 27–30. (In Chinese)
41. Xu, H.; Song, Y.P.; Goldsmith, Y.; Lang, Y.C. Meridional ITCZ shifts modulate tropical/subtropical Asian monsoon rainfall. *Sci. Bull.* **2019**, *64*, 1737–1739. [CrossRef]
42. Shiau, J.T. Fitting drought duration and severity with two-dimensional copulas. *Water Resour. Manag.* **2006**, *20*, 795–815. [CrossRef]
43. Chen, H.; Sun, J. Changes in drought characteristics over China using the standardized precipitation evapotranspiration index. *J. Clim.* **2015**, *28*, 5430–5447. [CrossRef]
44. Hamed, K.H. Exact distribution of the Mann-Kendall trend test statistic for persistent data. *J. Hydrol.* **2009**, *365*, 86–94. [CrossRef]
45. Sklar, A. Fonctions de répartition à n dimensions et leurs marges. *Publ. Inst. Statist. Univ. Paris* **1959**, *8*, 229–231.
46. Dong, L.Y. Application of Drought Index Based on Copula Function Model and FDA Method. Master's Thesis, Chongqing University, Chongqing, China, 2019.
47. Genest, C.; MacKay, J. The Joy of Copulas: Bivariate Distributions with Uniform Marginals. *Am. Stat.* **1986**, *40*, 280–283.
48. Cherubini, U.; Luciano, E. Bivariate option pricing with copulas. *Appl. Math. Finance* **2002**, *9*, 69–85. [CrossRef]
49. Zhao, H.Y.; Zhang, G.; Gao, G.; Lu, E. Characteristic analysis of agricultural drought disaster in China during1951–2007. *J. Nat. Disasters* **2010**, *19*, 201–206.
50. Chen, F.; Li, L.M. Researches on spatial and temporal succession law of agricultural drought in the past 60 years in China. *J. South China Normal Univ. Nat. Sci. Ed.* **2011**, *36*, 111–114.
51. Yao, Y.; Zheng, Y.G. The temporal and spatial characteristics of flood and drought during the recent 60 years in China. *Agric. Res. Arid Areas* **2017**, *35*, 228–232.
52. Zeng, Z.Q.; Wu, W.X.; Li, Y.M.; Zhou, Y.; Zhang, Z.; Zhang, S.; Guo, Y.; Huang, H.; Li, Z. Spatiotemporal Variations in Drought and Wetness from 1965 to 2017 in China. *Water* **2020**, *12*, 2097. [CrossRef]

 *atmosphere*

Article

Climate Variability, Dengue Vector Abundance and Dengue Fever Cases in Dhaka, Bangladesh: A Time-Series Study

Sabrina Islam [1], C. Emdad Haque [2,*], Shakhawat Hossain [3] and John Hanesiak [4]

[1] School of Health and Life Sciences, North South University, Dhaka 1229, Bangladesh; sabrina.islam@umanitoba.ca

[2] Natural Resources Institute, University of Manitoba, Winnipeg, MB R3T 2N2, Canada

[3] Department of Mathematics and Statistics, University of Winnipeg, Winnipeg, MB R3B 2E9, Canada; sh.hossain@uwinnipeg.ca

[4] Department of Environment and Geography, University of Manitoba, Winnipeg, MB R3T 2N2, Canada; john.hanesiak@umanitoba.ca

* Correspondence: cemdad.haque@umanitoba.ca

Abstract: Numerous studies on climate change and variability have revealed that these phenomena have noticeable influence on the epidemiology of dengue fever, and such relationships are complex due to the role of the vector—the *Aedes* mosquitoes. By undertaking a step-by-step approach, the present study examined the effects of climatic factors on vector abundance and subsequent effects on dengue cases of Dhaka city, Bangladesh. Here, we first analyzed the time-series of *Stegomyia* indices for *Aedes* mosquitoes in relation to temperature, rainfall and relative humidity for 2002–2013, and then in relation to reported dengue cases in Dhaka. These data were analyzed at three sequential stages using the generalized linear model (GLM) and generalized additive model (GAM). Results revealed strong evidence that an increase in *Aedes* abundance is associated with the rise in temperature, relative humidity, and rainfall during the monsoon months, that turns into subsequent increases in dengue incidence. Further we found that (i) the mean rainfall and the lag mean rainfall were significantly related to Container Index, and (ii) the Breteau Index was significantly related to the mean relative humidity and mean rainfall. The relationships of dengue cases with *Stegomyia* indices and with the mean relative humidity, and the lag mean rainfall were highly significant. In examining longitudinal (2001–2013) data, we found significant evidence of time lag between mean rainfall and dengue cases.

Keywords: climate variability; seasonality; dengue fever; vector; rainfall; Bangladesh

Citation: Islam, S.; Haque, C.E.; Hossain, S.; Hanesiak, J. Climate Variability, Dengue Vector Abundance and Dengue Fever Cases in Dhaka, Bangladesh: A Time-Series Study. *Atmosphere* **2021**, *12*, 905. https://doi.org/10.3390/atmos12070905

Academic Editors: Jong-Suk Kim, Nirajan Dhakal, Changhyun Jun and Taesam Lee

Received: 15 June 2021
Accepted: 12 July 2021
Published: 14 July 2021

Publisher's Note: MDPI stays neutral with regard to jurisdictional claims in published maps and institutional affiliations.

1. Introduction

The potential impacts of climate change on the human environment and infectious diseases are significant and alarming. Dengue/Severe Dengue Fever (DF/SDF) is one of the most rapidly growing arboviral diseases in the tropics, for which there is currently no universally accepted cure or vaccine. The rapid spread of both the dengue virus (DENV) and its mosquito vector (mostly *Aedes aegypti* and *Aedes albopictus*) in the past four decades poses an enormous risk to public health in tropical regions. The Halstead [1] and Gubler [2] studies suggested that the projected emergence will place around 2.5–3.0 billion people at risk of acute illness every year as tropical diseases spread to new areas, such as Europe and North America [3]. As well, *Aedes albopictus* plays a noticeable role in dengue transmission in the USA and Europe, whereas, in Asia, *Aedes aegypti* is more dominant in spreading dengue. Presently, populations of 129 countries worldwide are vulnerable with the risk of dengue infection—caused by both kinds of *Aedes* mosquitoes—of which, 70% of the actual disease burden exist in Asia. About 5.2 million dengue cases were recorded by the WHO in 2019 with an annual death count of 4032 people. Apart from dengue, these mosquitoes are also vectors of chikungunya, yellow fever and Zika viruses [4].

While some scholars account for climatic factors in their analysis of arboviral disease epidemiology [5,6], climatic factors and/or climate change are generally considered a discrete and separate entity in explaining disease dynamics. In clarifying dengue disease epidemiology, some quarters consider climatic aspects as unrelated factors [7]. Several investigations in Asian and Latin American countries, namely in Vietnam [8], Taiwan [9], and Ecuador [10], have confirmed a positive association between *Stegomyia* indices and *Aedes* abundance. We argue that climatic factors should therefore be considered as one of the principal determinants of the epidemiological complex that includes vector ecology, pathogen biology, disease transmission, disease occurrence and prevalence, and disease control, prevention, and cure. Such an improved understanding of emerging infectious diseases, including dengue, would enable us to more comprehensively map the process of disease occurrence and spread. This is especially vital for diseases like dengue for which risk assessment, prevention, and control are the only countermeasures available worldwide.

The relationship between climatic conditions and DF/SDF incidence is complex. A meta-analysis of the literature has revealed that rainfall, temperature and humidity are the most important explanatory variables in the transmission of dengue virus (DENV) through the means of vectors (*Aedes aegypti* and *Aedes albopictus*) and human hosts [11–13]. However, the specifics of these interactions vary widely from region to region and remain largely inconclusive in the current literature. In addition, several empirical studies in Asia and Latin America cautioned that the relationship between precipitation and dengue incidence may not be linear, as excess rainfall can negatively impact vector breeding [14,15]. Additionally, the *Aedes* mosquito—especially *Aedes aegypti*—is a type of vector that breeds in clean water and mostly found in different types and sizes of water containers. The role of artificial water containers, especially in urban areas, is therefore very important in *Aedes* mosquito breeding and dengue incidence. The artificial water containers, especially in the urban areas, thus play a pivotal role in dengue transmission through mosquito breeding, their life cycle, and by infecting people with DENV [16].

By addressing some of the major gaps in previous studies (elaborated on in the following section), our investigation attempts to make a novel contribution by considering the climate–vector–disease nexus in an integrated manner for understanding dengue transmission dynamics. It attempts to determine the effects of the main climate variables (temperature, relative humidity, rainfall, and seasonality) on dengue vector abundance and dengue disease occurrence in the city of Dhaka, Bangladesh. The specific objectives of the study are (1) to examine the relationship between the main climatic factors (temperature, relative humidity, and rainfall) and dengue vector abundance; (2) to examine the relationship between DF/SDF cases and vector abundance; (3) to examine the relationship between climatic factors and DF/SDF cases; (4) to map the patterns in seasonality and climate anomaly, along with their effects on DF/SDF cases.

This paper starts with an overview of the climate factors, specifically temperature, rainfall, relative humidity and dengue relationships, followed by an analysis of the trend in DF/SDF in Bangladesh and a critical review of the relevant studies in the country. The materials and methods are presented in Section 2, followed by the results in Section 3, and an analytical discussion in Section 4, with brief conclusions in Section 5.

1.1. Climate Factors and Dengue Relationships: An Overview

1.1.1. Temperature and Dengue

The relationship between DENV infection, DF and SDF incidence and temperature follows a complex, nonlinear trend. Most studies on dengue have been conducted in tropical areas where the annual temperature patterns are similar and disease transmission occurs at an optimal temperature range of 20–35 °C [17–19].

However, vector breeding and disease transmission are dependent on many other socioeconomic and human behavioral factors. It is also critical to distinguish between outside and ambient temperature when examining dengue vector breeding conditions. For example, in sub-tropical regions, breeding can still occur at sub-optimal outside temperatures (i.e., during winter) if the ambient (indoor) temperature is higher and heating and standing water are available [17]. Conversely, even if the outside temperature is optimal, if the ambient temperature in buildings is lower due to air conditioning, the vector may not breed or multiply sufficiently to cause an outbreak. The breeding patterns of *Ae. aegypti*—the major vector for dengue—may not necessarily correlate with outdoor temperature, rather they may be associated with ambient temperature.

1.1.2. Rainfall and Dengue

The relationship between dengue incidence and rainfall is dependent on numerous complex, interlinked factors. The dengue epidemiology literature reveals that dengue outbreaks in most countries coincide with the wet season and increased precipitation in general [12,13,19]. In this regard, Kuno [17] noticed a positive association between rainfall and larval density and dengue incidence that has since been documented in many tropical countries. However, this causal pathway cannot be universally generalized, as dengue outbreaks follow different climatic patterns in certain regions. Moreover, excess rainfall can negatively impact vector breeding [14,15] by washing off the vector breeding sites and thus can affect dengue outbreaks.

1.1.3. Relative Humidity and Dengue

Despite great interest within the research community in the association between climatic factors and dengue incidence, research on relative humidity as an important climatic factor has been relatively scant. Furthermore, the results of the few studies have also been inconsistent and inconclusive. An Indonesian study revealed that the most important predictive factor for dengue outbreaks in that country was relative humidity, with a 3–4-month lag time [14]. This research revealed that low relative humidity during September and October is usually followed by a dengue outbreak early the following year. It is thus highly probable that if seasonal conditions (average temperature and humidity) are shifted due to climate change, seasonal incidences of dengue would be shifted as well.

1.1.4. Dengue Studies in Bangladesh

Bangladesh is situated in the tropical monsoon climate zone. Dhaka—the capital city—and other major urban centers experience a hot, wet and humid tropical climate. Bangladesh has a country-wide monsoon mean temperature of about 29 °C [20], which falls within the optimal range for both mosquito breeding and dengue transmission [17–19]. The major cities of Bangladesh (a country with a population of about 160 million in an area of 143,000 km^2) have experienced a major resurgence of dengue since 2000 [21,22].

The first epidemic of SDF occurred in 2000 in the cities of Dhaka, Chittagong and Khulna. During this dengue epidemic, a total of 5551 infections were reported and 93 patients died [23]. Since then, serious concern has been expressed regarding the lack of understanding of the dynamics of dengue transmission and the urgent need for improved disease management. According to Sharmin et al. [24], the 2000 outbreak resulted from a virus strain originating in Thailand, located to the east of the country. They also added that dengue cases have remained underreported in Bangladesh as rural people only visit hospitals in the most severe cases. Rahman et al.'s [25] research indicated some degree of correlation between the DF/SDF outbreak in 2000 and monsoon seasonal conditions as the outbreak started in late June 2000, peaked in September (during the rainy season) and decreased in the dry winter season of the same year.

DF/SDF or similar fever is not a new disease in the country. For example, Hossain et al. [22], after analyzing samples from febrile patients between 1996 and 1997, suggested that dengue transmission was ongoing in the country well before 1996. The Sharmin et al. [24] and Morales et al. [26] studies further noted that dengue could be traced back to 1964 in Bangladesh (then East Pakistan)—much earlier than the major outbreak in 2000.

Mortality rates have decreased significantly since the outbreak of 2000; however, a sizeable population is still infected with DENV every year (Figure 1). The distribution of dengue cases and deaths over the period of 2001–2019 in Dhaka is illustrated in Figure 1.

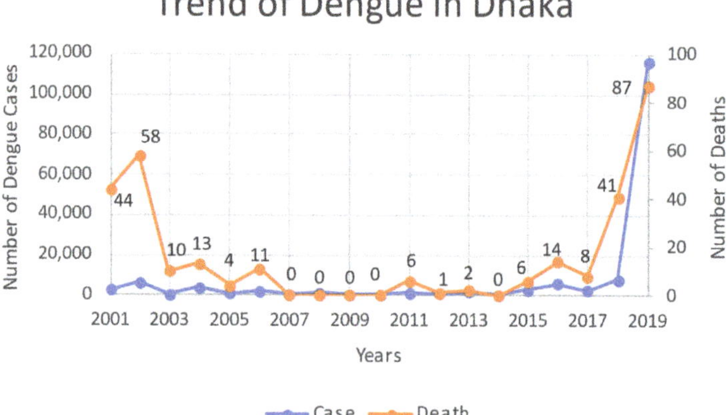

Figure 1. Distribution of dengue cases and number of deaths in the city of Dhaka, Bangladesh, 2001–2019. (Source: Data procured from the Directorate General of Health Services (DGHS), Bangladesh, 2001–2013; Dhaka Tribune [27]; Shirin et al. [28]).

Due to data unavailability for the year 2000 for Dhaka, both the reported cases and number of fatalities were excluded. Though the dengue incidences have decreased over the years until 2015, it has followed cyclical and fluctuating patterns since 2001. It is evident that the number of deaths has decreased dramatically—from 44 deaths in 2001 and 58 deaths in 2002 to no deaths in 2014. (Figure 1). It then started increasing, resulting in 41 deaths in 2018 and an estimated number of 87 deaths in 2019. There was a large increase in reported dengue cases and number of deaths in 2019. A total of 80,040 dengue cases were reported officially between 1 January and 13 September 2019 with a total of 60 deaths due to SDF and shock syndrome [27]. According to the existing literature, all four DENV serotypes prevail in the city of Dhaka, with DENV-3 being the dominant one [29]. During the 2002 major outbreak, DEN-3 predominated, and subsequently other serotypes were also found to be in circulation.

Studies on the relationship between climate and dengue transmission in Bangladesh have primarily focused on patients in hospitals and clinics [22,24,25,30], and only a few investigations have hitherto been carried out in the country [31–33]. Paul et al. [34] studied the effects of climatic factors on *Aedes* abundance in the city of Dhaka, Bangladesh, limiting their study only to climate factors and vector abundance relationships. They concluded that rainfall, temperature, and relative humidity significantly affected the mean abundance of mosquitoes.

In a rare study on climate–dengue case relationships in Bangladesh, Hashizume et al. [31] conducted a time-series analysis of the trend between hydro-climatological variability and DF cases, and found a positive association between DF cases with high as well as very low river levels with varying weekly time lags of 0–19 weeks. Islam et al.'s [35] recent study of the city of Dhaka inferred that dengue incidence is significantly associated with the monthly mean temperature, total rainfall, and mean humidity. The study established a linear relationship of the climatic factors and the dengue incidence, while not accounting for the aspects of seasonality and the vector relationships.

Karim et al. [32] examined the influence of climatic factors on dengue cases in the city of Dhaka, and found that rainfall, maximum temperature, and relative humidity could explain 61% of the variability in reported dengue cases with a two-month lag period. The study revealed that the arrival of the monsoon season, with a peak in August, was sufficient to explain most of the reported dengue cases. However, the role of vector mosquitoes in dengue transmission was not considered. An investigation in the city of Dhaka revealed that both *Ae. aegypti* and *Ae. albopictus* larval populations peaked in July at the height of the monsoon [36].

In studies on dengue in Bangladesh, we argue that the focus has generally been on bivariate relationships either between climate factors and vector abundance or between vector abundance and dengue cases [24,32,33]. In Bangladesh studies, the climate–vector–disease nexus under one research framework has not yet been explored. We therefore assert that considerable gaps still exist in our understanding of the complex climate–vector–disease nexus and how these relationships are being affected by confounding factors such as urbanization and human behavior. As there is not yet a cure or universally available vaccine for dengue, it is vital to improve our understanding of risk factors in order to effectively control and prevent the spread of the disease in developing countries like Bangladesh.

2. Materials and Methods

2.1. Study Area and Design

The city of Dhaka is the largest urban center of Bangladesh. Considering its socioeconomic, political and demographic significance, pivotal standing in terms of population health risk to infectious diseases, and recurring number of dengue cases, the Dhaka City Corporation (DCC) was chosen as the study area for this investigation (Figure 2). Located on the banks of the Buriganga River, Dhaka has an area of 126.34 square kilometers (census 2011) [37] and ranks 11th among global mega-cities with a population of 18.2 million [38]. The city experiences a hot, wet, and humid tropical climate and a distinct monsoon season, with an annual average temperature of 28 °C (82 °F) and monthly means varying between 20 °C (68 °F) in January and 32 °C (90 °F) in May. Nearly 80% of the annual average rainfall of 1854 mm (73 in) occurs between May and September [39].

2.2. Data Collection Techniques

First, the meteorological data required for the study included temperature, relative humidity and rainfall on daily, monthly and yearly time scales. These data were obtained from the Bangladesh Meteorological Department (BMD) in Dhaka for the 1985–2014 period. A near standard 29-year period (1985–2014) was used as the climate baseline to calculate climate anomalies in relation to dengue cases [40]. We could use only 29 years of data instead of 30 years as data prior to 1985 were not available from the BMD. The data collection method was different before 1985 and therefore data were not compatible with data available from 1985 onwards. The data were obtained from a single observatory located at the Dhaka Airport (Old) (Figure 2). Monthly averages were calculated from the daily data for trend and seasonality analysis of the selected variables (temperature, rainfall and humidity) in relation to dengue cases.

Notably, we could not use daily data since daily data were not available for the *Aedes* mosquito or dengue cases. The data were homogeneous according to the BMD for the period mentioned. Second, the data for the House Index (HI), Breteau Index (BI) and the Container Index (CI) for *Ae. aegypti* larvae for the city of Dhaka were obtained from the Directorate General of Health and Services (DGHS) of the Government of Bangladesh for the 2002–2013 period. There were two limitations: (i) continuous time series data were not available as entomological surveys were conducted with interruptions, and (ii) there were common as well as uncommon surveyed areas in the sequential surveys. Therefore, the larval data from entomological surveys were available only for seven years: 2002, 2003, 2004, 2005, 2009, 2012 and 2013. These surveys were carried out during the same period (monsoon season) of each year, namely during June–October months.

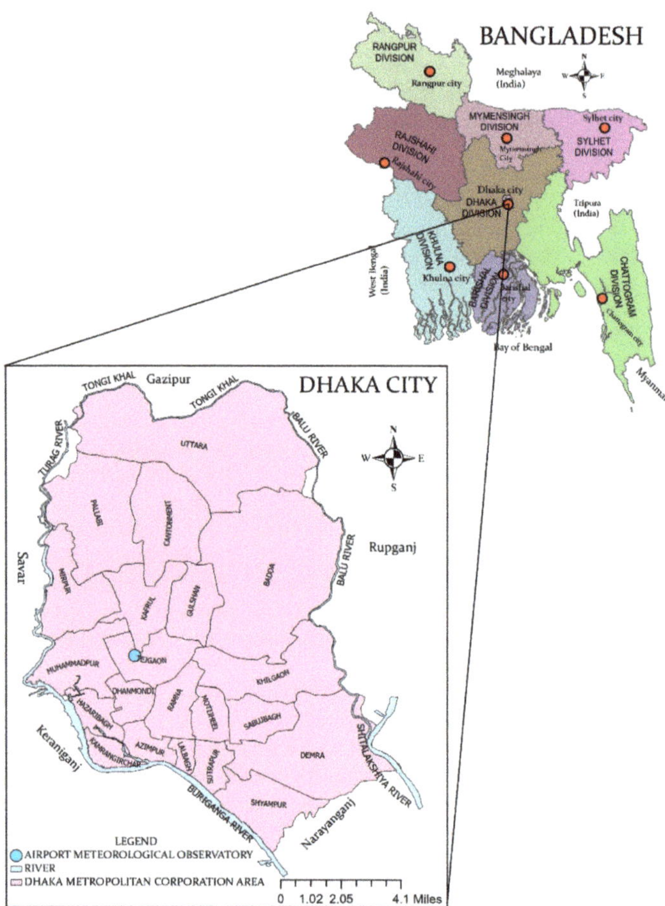

Figure 2. Location map of the study area—the City of Dhaka, Bangladesh.

However, as noted above, the geographic areas surveyed by the DGHS had common as well as varied areas from year to year. In order to maintain consistency, ensure comparability and reduce possible biases, we selected only the common areas within which data collection was repeated for at least five years. We re-calculated all indices (HI, BI and CI) for the city of Dhaka based on DGHS datasets. As no data were available for 2006, 2007, 2008, 2010, and 2011 from any sources in Bangladesh to conduct a field-data based time-series analysis, these missing data were denoted as "missing at random". Following Little and Rubin [41] and Weerasinghe [42], these missing data were imputed by applying the Spline Interpolation Method (see R-package "imputTS") [43,44] and the regression imputation method [45]. Third, the dengue case data (2002–2013) were also obtained from the DGHS. All statistical analyses were performed using Microsoft Excel and the statistical software R [46].

2.3. Statistical Analyses

We applied various statistical techniques to identify interactions within the climate–vector–disease nexus. To the best of our knowledge, this relationship has not been previously studied to a significant extent, especially in the South Asian context, and thus the constituent dynamics are not understood well. Our initial approach is to analyze the effects of climatic variables on vector abundance, then subsequently analyze the effect of vector abundance on dengue incidence, conceptualizing this pathway as a step-by-step process. In addition, we also analyzed seasonality factors that may affect disease patterns over the years. To implement the above, the following data analyses were performed based on the collected meteorological, entomological and dengue case data:

Analysis of climate factors vs. vector indices: first, we calculated the monthly mean temperature (MT), mean relative humidity (MH), mean rainfall (MR), and lag mean rainfall (LMR) of one-month lag. We then attempted to find relationships between each of the indices (CI, BI and HI) and the climate factors (MT, MH, MR, and LMR) for the period of 2002–2013. Linear regression model assumes a fixed parametric form of the relationship between vector indices and climate factors. The generalized additive model (GAM) [47] does not assume any specific form of this relationship and can be used to reveal and estimate nonlinear effects of the climate factors on the vector indices. To implement this relation, the following GAM was used:

$$\text{Index}_{it} = \beta_0 + f(MT_t) + f(MH_t) + f(MR_t) + f(LMR_t) + \text{error}_t \tag{1}$$

where β_0 is the intercept; i is HI, CI or BI; t is time; f is the unknown smooth functions of climatic factors which are determined by the data.

We used an autocorrelation of lag 1 to measure the relationship between one month's temperature (Y_t) and the previous month's temperature (Y_{t-1}) (same for humidity, rainfall and lag mean rainfall). The values of the autocorrelation function (ACF) and partial autocorrelation function (ACF) helped us to identify the autoregressive order and moving average order, respectively. Next, we estimated the GAM model parameters after adjusting the order in the estimation procedure.

Analysis of vector abundance vs. dengue case incidence: here, we emphasized to examine the relationship between vector abundance (by using HI/BI/CI) and dengue cases for the period of 2002–2013. Since our response (dengue cases) was counts, the following Poisson regression model was used to model vector abundance vs. dengue case incidence:

$$\log(\mu_t) = \beta_0 + \beta_1 HI_t + \beta_2 BI_t + \beta_3\, CI_t \tag{2}$$

where μ_t is the mean case count with respect to time t.

Analysis of climate factors vs. dengue case incidence: we examined the relationship between dengue cases and climatic factors for specific months over the 12-year study period (2002–2013), using the Poisson regression model. In statistics, when modeling count data (number of dengue cases, in our study), the Poisson regression model is used.

The relationship among the climate–vector–dengue nexus was established based on the above modeling approaches. However, we could not consider seasonality effects in this analysis because appropriate vector data for seasonality were unavailable. For seasonality analysis, we used a different dataset which has monthly average temperature, rainfall and relative humidity data for the period of 2001–2013. A relationship of climatic factors and dengue cases was established after adjusting the seasonality effect.

Analysis of seasonality vs. dengue cases: dengue case data for Dhaka city were plotted against the climate data variables (temperature, rainfall and relative humidity) from the BMD for the period of 2001–2013 in order to map seasonal variations in dengue cases. We then determined the relationship between dengue cases (which is again count response) and climatic factors (MT, MH, and MR), using the following Poisson regression model:

$$\text{Log}(\mu_t) = \beta_0 + \beta_1 MT_t + \beta_2 MH_t + \beta_3\, MR_t \tag{3}$$

where μ_t is the mean case count with respect to time t. After establishing the climate–vector–dengue nexus for specific months over 2002–2013 and the seasonal variation of dengue while determining the effects of climatic factors on dengue, it is important to examine the variation in dengue cases based on climate anomalies (of annual temperature and rainfall over the period of 2001–2013). To implement this, we calculated the annual mean of temperature and rainfall for the period 1985–2013 to act as the climate normal baseline for the study. The annual anomaly was found by subtracting the climate mean of temperature, relative humidity and rainfall from the data for each individual year. This can be expressed as:

$$\text{annual anomaly}_{(2001\text{-}2013)} = \text{annual mean}_{(2001\text{-}2013)} - \text{climate normal}_{(1985\text{-}2013)} \quad (4)$$

The data obtained from relation (4) were then plotted to show the annual anomaly over the 2001–2013 period. This depicted the changes in dengue disease occurrence in association with the annual anomaly over the years for the 2001–2013 period.

Notably, prior to a major dengue outbreak in 2000, case data collection by the public health agencies was sporadic and limited to clinical data. Large scale (population-based) dengue case data collection began only after the outbreak in 2000. Subsequent to the 2000 dengue outbreak, large scale entomological data collection was also initiated and were available from 2002 onward. Limited by data unavailability, we used the 13-year period (2001–2013) for examining the relationships between seasonality and dengue cases, and a 12-year period (2002–2013) for examining the relationship between entomological data-based vector abundance with dengue cases, and the climate variables. These time frames are graphically shown in Figure 3 for clarity.

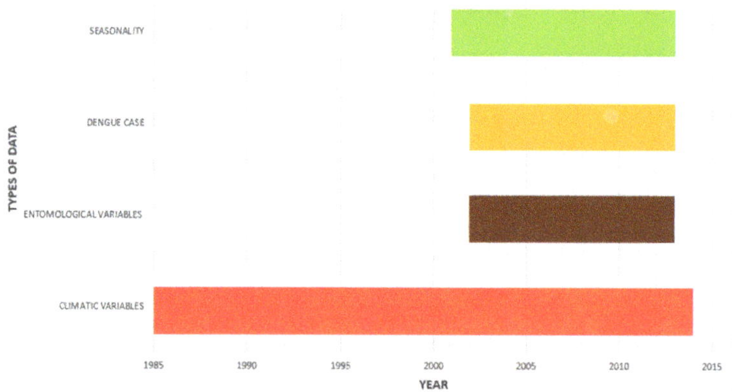

Figure 3. Overlapping time periods of different datasets.

This research was approved by the Bangladesh Medical Research Council (Bangladesh) and the Joint Faculty Research Ethics Board of the University of Manitoba (Canada). Administrative permission was granted by the Government of Bangladesh to access and use the meteorological data from the Bangladesh Meteorological Department, and entomological and dengue case data from the Directorate of Public Health and Services.

3. Results

In this section we provide the data analysis results based on the sequence of analyses discussed in the previous section.

3.1. Analysis of Climate Factors vs. Vector Indices

We examined the effects of climatic factors on each of the *Stegomyia* indices (CI, BI and HI) for the city of Dhaka over the 2002–2013 period, based on observed (by BMD and DGHS) and our imputed meteorological and entomological data (Table 1).

Table 1. Vector indices and climatic variables for specific months (2002–2013) in the city of Dhaka, Bangladesh.

Years	Months	%CI	%BI	%HI	MT (°C)	MH (%)	MR (mm)	LMR (mm)
2002	Aug.	7.49	15.28	14.25	28.6	81	8.77	14.39
2003	Aug.	16.31	16.84	8.74	29.4	78	6.52	6.16
2004	Jun.	29.91	32.54	14.04	28.5	81	15.87	5.23
2005	Sep.	15.7	13.91	10.47	28.9	81	17.13	11.65
2006	Aug.	8.11	14.69	8.49	29.1	77	5.39	10.68
2007	Aug.	13.61	39.14	10.73	29.1	80	16.29	24.29
2008	Aug.	26.46	71.07	15.58	28.8	81	10.29	18.16
2009	Jun.	40.91	94.29	21.43	30.2	74	5.67	5.42
2010	Aug.	51.56	96.59	26.65	29.5	78	10.97	5.39
2011	Aug.	54.36	81.7	29.45	28.5	82	13.19	11.48
2012	Sep.	45.6	57.33	28.01	29	79	2.70	9.10
2013	Oct.	21.47	30	20.53	27.2	78	4.23	5.73

Note: The specific months in this study were used because data were available only for these months as collected and provided by the DGHS.

Results of our univariate analysis revealed that there is a nonlinear relationship between CI and mean rainfall (MR) or lag mean rainfall (LMR). We used a smoothing method based on cubic splines to estimate the functional effect of climate factors on the vector indices GAM (1).

The results of the fitted model (1) showed that neither mean temperature (MT) nor mean relative humidity (MH) were significantly associated with CI. However, MR was found to be positively associated with CI (Table 2). The adjusted $R^2 = 0.79$ means that the fitted model can explain 79% of the variability in CI due to MR.

Table 2. Association between the *Stegomyia* indices and climate variables, dengue cases with the *Stegomyia* indices and dengue cases with climate variables.

Association of *Stegomyia* Indices with Climate Variables											
	Association with CI				Association with BI				Association with HI		
Variables	MT	MH	MR	LMR	MT	MH	MR	LMR	MT	MH	MR	LMR
p-value	0.294	0.360	0.0372 *	0.0762	0.7976	0.0208 *	0.0186 *	0.8572	0.448	0.259	0.197	0.678
R^2 (adj)	0.79				0.72							
Association of dengue cases with *Stegomyia* Indices											
Variables	HI	BI	CI									
p-value	0.005 *	<0.001 *	<0.001 *									
R^2 (adj)	0.49											
Association of dengue cases with climate variables											
Variables	LMR	MH										
p-value	0.0279 *	0.0463 *										
R^2 (adj)	0.93											

Note: * significant at 0.05. HI = No. of positive HHs/no. of HHs visited, CI = No. of positive containers/No. of wet containers, BI = No. of positive containers/No. of HHs visited. MR = mean rainfall, MH = mean humidity, LMR = lag mean rainfall for one month.

To check the normality of the error term in the fitted model (1), we drew the autocorrelation (ACF) and partial autocorrelation plots (PACF) of residuals. The PACF plot (Figure S1) indicated that the residuals in the fitted model were uncorrelated with mean zero and constant variance.

We fitted the same model (1) when the response was BI. Both MT and LMR showed positive relationships with BI but were not significant. However, the results indicated that BI was significantly related to MH and MR (Table 2). The MH was negatively related to BI, meaning that as MH increases, BI—and thus vector density—decreases. The adjusted R^2 value = 0.72 means that the fitted model can explain 72% of the variability in BI associated with MH (Table 2). We checked the model assumption using ACF and PACF plots, which revealed that the residuals were uncorrelated with mean zero and constant variance (Figure S2).

We fitted the same model (1) again when the response was HI, revealing that there were positive relationships between HI and MT, MR and LMR, and a negative relationship with MH. None of the climatic factors were significantly related to HI (Table 2). The above findings reveal how dengue vector abundance and distribution are impacted by temperature, rainfall, and relative humidity. As the vector is a means rather than a final phenomenon in terms of dengue disease, it is critical to extend our examination to the relationship between dengue and the *Stegomyia* indices.

3.2. Analysis of Vector Abundance vs. Dengue Case Incidence

We fitted the Poisson regression model (2) of dengue cases with each of the indices HI, BI, and CI. The results indicated that there were highly significant relationships between dengue cases and the indices (HI, BI and CI) (Table 2). It is evident from the results that both BI and CI were highly significant variables among the three indices. The adjusted R^2 value of 0.489 showed that the model could explain about 49% of the variability in dengue cases accounted for by the indices (Table 2). We examined the ACF and PACF plots of standardized Pearson residuals of the fitted model (2), which revealed that the residuals behaved as uncorrelated with mean zero and constant variance.

3.3. Analysis of Climate Factors vs. Dengue Case Incidence

We fitted the Poisson regression model (2) of dengue cases with each of the climatic factors MT, MH, MR, and LMR. We present below the results from the estimated model, which reveal the relationships between dengue cases and climatic factors for the study period (i.e., 2002–2013) for specific monsoon months, encompassing June to August.

Results from the fitted model (2) indicate that there are significant relationships between dengue cases and each of MH, and LMR (Table 2). It is evident from these results that MH and LMR are significantly related to dengue cases. The adjusted $R^2 = 0.93$ means that the model can explain about 93% of the variability in dengue cases resulting from climatic factors (Table 2). We examined the ACF and PACF plots of standardized Pearson residuals for the fitted model (2), which indicated that the residuals behaved as uncorrelated with mean zero and constant variance.

3.4. Analysis of Seasonality vs. Dengue Cases

After analyzing the relationship among the climate–vector–dengue nexus above based on the available DGHS data only during monsoon season, we considered another dataset; because the previous dataset only had data for monsoon seasons, and for seasonality analysis, we needed monthly data for all the climatic variables for the period considered under the study. This was used to analyze seasonal variability of dengue cases versus monthly average temperature, rainfall and relative humidity for the 2001–2013 period. Here, we analyzed the seasonality in terms of dengue cases only and determined the relationship of climatic factors and dengue cases while taking the seasonality in account.

First, the monthly time series of temperature, relative humidity and rainfall data along with dengue cases were plotted in Figure 4 to view the patterns in their distribution over a 13-year period (2001–2013). Figure 4a indicates that dengue cases follow a peak returning pattern, with a sharp rise (of more than 3000) in the second year of the study (2002) and peak every other year until 2008, after which the peaks occurred every two years. Figure 4b reveals that temperature followed a similar pattern over the same 13-year period, with a peak of 30 °C (monthly average temperature) during months of July/August of most of the years and a low of 16 °C (monthly average temperature) in early 2003. Figure 4c shows the monthly average relative humidity pattern over the 13-year study period, which was cyclical with a consistent yearly peak of 80–85%. Finally, Figure 4d also depicts a yearly cyclical and seasonal pattern of rainfall. The average monthly rainfall for the aforementioned period varies between 100 and 150 mm with occasional departures; for example, the peaks in 2004, 2006, 2007 and 2009 with highest peak in 2004 having more than 250 mm monthly average rainfall.

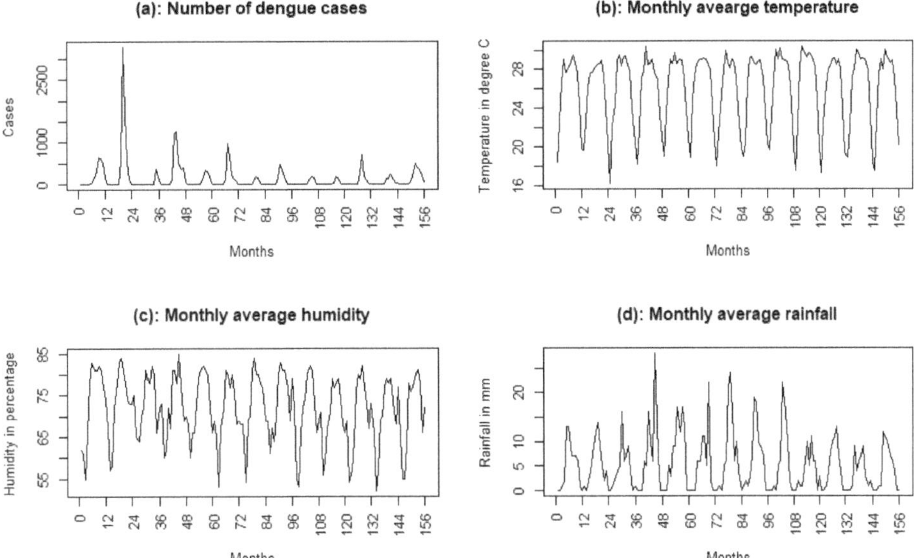

Figure 4. Patterns in the distribution of monthly data for dengue cases and selected climate factors for the 2001–2013 period in the city of Dhaka, Bangladesh: (**a**) number of dengue cases; (**b**) average monthly temperature; (**c**) average monthly relative humidity; (**d**) average monthly rainfall. **Note:** Every 12 months = 1 year, starting from year 2001 to year 2013. Thus, months 0–12 = year 2001, months 12–24 = year 2002, months 24–36 = year 2003, months 36–48 = year 2004, months 48–60 = year 2005, months 60–72 = year 2006, months 72–84 = year 2007, months 84–96 = year 2008, months 96–108 = year 2009, months 108–120 = year 2010, months 120–132 = year 2011, months 132–144 = year 2012, months 144–156 = year 2013.

Overall, all of the climatic factors show a yearly cyclical pattern in average monthly temperature, relative humidity, and rainfall, as shown in Figure 4b–d, respectively. Figure 4a shows that dengue cases also follow a seasonal pattern, with incidences being highest during the monsoon season and lowest during the pre-monsoon season (January–April) in alternating years.

The autocorrelation and partial autocorrelation plots of dengue cases, monthly average temperature, monthly average relative humidity and monthly average rainfall for the 13-year period are shown in Figure S3. ACF gives us values of autocorrelation of any series (such as, temperature) with its lagged values. As depicted in Figure S3 there is a significant correlation at lags 1 and 2 followed by correlations that are not significant. This pattern indicates a moving average (MA) process of order 2. Other Figures clearly indicate the seasonal behavior of the monthly temperature, relative humidity and rainfall.

Second, when we checked the ACF and PACF plots of the residuals for the 13-year period to determine seasonal variations in the climatic parameters (Figure S4), we found an autoregressive pattern of order 1. To account for this, we fitted the Poisson regression model (3) using the R-function *tsglm* from the *tscount* R-package.

From the fitted model output, the estimated coefficient and its standard error for temperature were found to be 0.10 and 0.003, respectively. The confidence interval for the coefficient of temperature is (0.10, 0.11) which did not include zero. This means that the temperature is significant at a 5% level; that is, temperature plays a significant role in the number of dengue cases in the city of Dhaka. Since the estimate is positive, the number of dengue cases increases as the temperature increases, while adjusting for relative humidity and rainfall. For relative humidity, the estimate is 0.23 and the standard error is 0.002. The confidence interval for the coefficient of relative humidity is (0.22, 0.24) which did not include zero. This means that relative humidity is also significant at the 5% level and plays a significant role in the number of dengue cases in the city of Dhaka. As the estimate is positive, the number of dengue cases increases as relative humidity increases, while adjusting for temperature and rainfall. For rainfall, the estimate is −0.10 and the standard error is 0.002. The confidence interval for the coefficient of rainfall is (−0.10, −0.09) which does not include zero. This means that rainfall is significant at the 5% level; that is, rainfall plays a significant role in the number of dengue cases in the city of Dhaka. As the estimate was negative, the dengue cases decrease as the rainfall increases, while adjusting for temperature and relative humidity.

3.5. Analysis of Climate Anomaly vs. Dengue Cases

After the seasonality analysis for dengue cases, we aspired to test the effects of climate anomalies on dengue cases. For this, we calculated the temperature anomaly using the relation (4) and plotted it in Figure 5 against dengue cases for the period of 2001–2013.

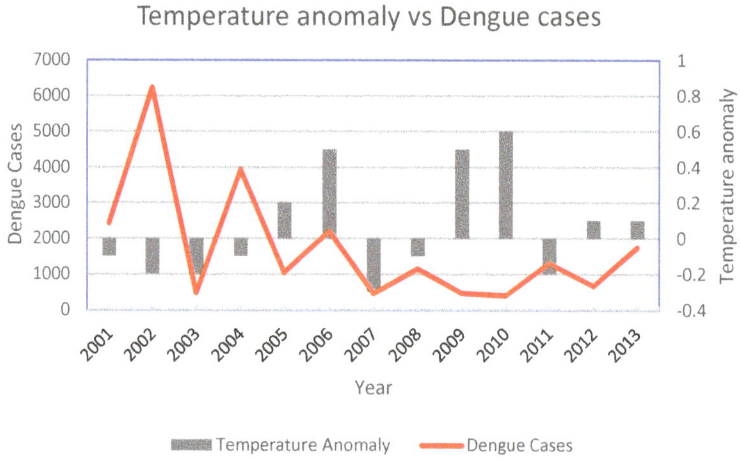

Figure 5. Distribution of temperature anomaly vs. dengue cases in the city of Dhaka, Bangladesh for the 13-year study period (2001–2013).

As shown in Figure 5, when the temperature anomaly was −0.1 °C (average yearly temperature 25.9 °C for years 2001, 2004 and 2008), or the anomaly was −0.2 °C (average yearly temperature 25.8 °C in 2002, 2003 and 2011), the number of dengue cases was higher than the number of dengue cases when the climate average temperature (26 °C) (1985–2013) occurs.

We also observed that at higher anomalies of 0.1–0.6° C (temperatures rise to 26.1–26.6 °C in years 2005, 2006, 2009, 2010, 2012 and 2013), the dengue cases were lower than the dengue cases in an average climate temperature (26 °C). Thus, the number of dengue cases in the city of Dhaka declined when the average yearly temperature was 26 °C and higher.

We calculated the rainfall anomaly using the relation (4) and plotted it against dengue cases in Figure 6 for the 2001–2013 period. The dengue cases tended to be higher for the years with a rainfall anomaly of −38.18 (annual average rainfall of 133 mm in 2013) and −10.76 anomaly (160 mm rainfall in 2006) as compared with the number of dengue cases for the average rainfall of 170 mm during the 29-year period of 1985–2013.

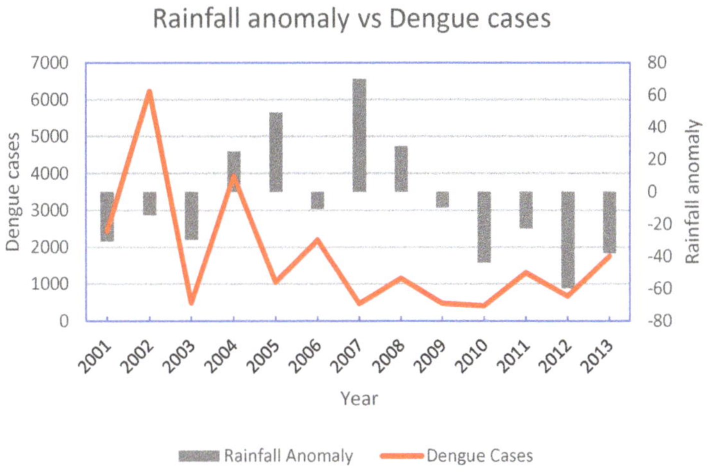

Figure 6. Distribution of rainfall anomaly vs. dengue cases in the city of Dhaka, Bangladesh for the 13-year study period (2001–2013).

We also observed that at higher rainfall anomalies, such as +69.74 in 2007 (rainfall of 240 mm), the dengue cases tended to be lower than the number of dengue cases for the average rainfall. Based on the data collected, we infer that with an increase in the amount of rainfall (more than annual average of 200 mm), the number of dengue cases tend to decline.

4. Discussion

Departing from conventional bivariate analytical approaches, in this study we undertook a step-by-step approach to determine the relationships between climatic factors (i.e., temperature, relative humidity and rainfall) and *Stegomyia* indices [9,10,48,49], and subsequently between *Stegomyia* indices and dengue cases in the city of Dhaka, Bangladesh. Besides analyzing the effects of seasonal variability and climate anomalies on dengue cases, our study sought to develop a more complete understanding of the effects of climate on dengue cases by viewing the climate–vector–disease nexus as a sequential, step-by-step process.

In designing our study, we drew from Githeko's [50] study in Bangladesh. Githeko [50] observed that most previous studies on dengue only considered climatic or biological effects in isolation and emphasized that since climatic factors directly influence the breeding and prevalence of dengue vector mosquitos (e.g., *Ae. aegypti*), it was vital to study the effect of climate on the biological domain in order to fully understand the dynamics of dengue outbreaks. Following this recommendation, we attempted to determine the relationships within the climate–vector–dengue nexus, with the aim of meeting the objectives of this study.

In regard to the first objective, the results of our study provide evidence that overall dengue vector abundance is significantly influenced by climatic factors, though the significance of this relationship varies for each index and pair of factors. We found that mean rainfall (MR) and lag mean rainfall (LMR) at one month were positively and significantly related to Container Index (CI), i.e., an overwhelming majority (i.e., 79%) of the variability in CI could be explained by MR and LMR. The mean rainfall contributes to the development of the *Ae. aegypti* when the daily or weekly mean rainfall is adequate for creating the breeding site as well as retaining the water for the growth of *Ae. aegypti* larvae and pupae. The lag mean rainfall indicates how much time needs to be passed after the actual rainfall to create the breeding site and having actual larval and pupal growth in the site. The mean relative humidity (MH) and mean rainfall (MR) were also found to be significantly related to the Breteau Index (BI), although MH was negatively associated with BI.

The regression model reveals that climatic factors account for about 72% of the variability in BI, which aligns with conclusions drawn by several other studies conducted in tropical regions [8,9,13,48,51]. However, while previous studies made the link between rainfall, water levels in indoor and outdoor containers, and the consequent increase in vector breeding and density and dengue prevalence, our findings and that of several other studies revealed that the dynamics of dengue spread are not so straightforward, and that the effects of temperature and relative humidity were also significant and complex.

Several comparable studies in Vietnam, Ecuador, Taiwan, Thailand, and China, which investigated the relationships between *Stegomyia* indices and *Aedes* abundance, provide useful insight into the impact of weather conditions on mosquito ecology. Pham et al.'s [8] study in the central high province of Vietnam confirmed a positive association between HI, BI, and CI and elevated temperatures, high humidity and rainfall, and a negative association with hours of sunlight [8]. Stewart-Ibarra et al. [10], in an empirical investigation of *Ae. aegypti* in Ecuador, confirmed that mosquito oviposition (egg-laying) was significantly driven by rainfall and minimum temperature. Tseng et al. [9] in their study of Taiwan found that temperature, lagged rainfall, and lagged density levels had positive and significant effects on the density of the mosquito population. Similar to these studies, the results of the present study in the city of Dhaka, Bangladesh revealed positive and significant relation of *Stegomyia* indices and the *Aedes* abundance.

Both Nakhapakorn and Tripathi [52] and Naish et al. [13] found that in Thailand, humidity is a crucial variable in the spread of dengue as high humidity—along with high temperatures and the presence of stagnant water—creates ideal breeding conditions for *Ae. aegypti*. Cheng et al. [51] examined the climate–vector abundance relationship in terms of the risk of dengue outbreak in Guangzhou, China, and found that that high precipitation during the monsoon increases vector abundance, but early and more frequent intervention and less vertical transmission can reduce the risk of dengue outbreak in successive time periods. Our study also corroborates these inferences by showing significant relationship between *Stegomyia* indices and the climate variables.

The second objective of our study was to determine the relationship between DF/SDF cases and vector abundance, as established by the parameters of the *Stegomyia* indices—a common epidemiological research framework that has been widely used in tropical areas [9,10,48,49]. Our empirical investigation in the city of Dhaka, Bangladesh has revealed that dengue cases in the city are significantly related to *Stegomyia* indices. We also found that CI is the most significant index, associated with dengue incidence, accounting for 49% of the variability in dengue cases.

In the context of Dhaka, these findings underscore the pivotal role of rainfall and stagnant water, relative to temperature and relative humidity, in affecting vector abundance and dengue incidence, as CI reflects the percentage of water-holding containers infected with larvae and/or pupae. These results are supported by Stewart-Ibarra et al.'s [10] study in Ecuador, that found mosquito oviposition was significantly driven by rainfall and minimum temperature, and that areas with large numbers of water storage containers played a major role for pupal development—the single most significant predictor of dengue outbreaks.

Previous studies on vector–dengue incidence have put forward two main philosophies regarding disease prevention and management. One school of thought argues that since dengue incidence is mainly driven by higher vector breeding associated with high temperature and rainfall, the policy focus should be on vector control. Pham et al. [8], for example, observed that in Vietnam the incidence of dengue fever is significantly and positively associated with *Stegomyia* indices. They argued that because higher vector abundance leads to higher rates of dengue infection, it is of utmost importance to control mosquitos during periods of high temperature and rainfall to reduce the risk of any outbreak. Similarly, the Barrera, Amador and MacKay's [48] study in San Juan, Puerto Rico, asserted that although both weather and anthropogenic activities are responsible for the abundance of *Ae. aegypti*, it is oviposition which is significantly correlated with dengue incidence, and which requires immediate attention.

Another school of thought asserts that *Stegomyia* indices, especially larval indices, are not sufficient to explain the vector–disease incidence relationship. Bowman et al. [49] reviewed the evidence systematically and concluded that the association between *Stegomyia* indices and dengue transmission was insufficient to predict an outbreak and called for the use of standardized study design and routine adult mosquito sampling in order to better understand vector ecology. In their study of the city of Kaohsiung, Taiwan, Chang et al. [53] came to a similar conclusion and argued that there may be different outbreak thresholds for different regions, local ecologies and herd immunity levels. As our findings revealed the relationship between vector abundance and dengue cases to account for only 49% of the variability in dengue incidence in the city of Dhaka, this study leaves room for other factors such as differences in adult mosquito numbers, infection rate of the adult mosquito, extrinsic incubation period, herd immunity, and population density, to be integrated into future epidemiological models for dengue.

The third objective of our study was to determine the relationship between climatic factors and DF/SDF cases. As the majority of studies on the climatic dimensions of dengue emergence ignored the entomological dimension, we pursued a longitudinal study of the city of Dhaka in order to more fully understand the disease propagation complex. The results of our generalized linear model reveal that, when the relationships between dengue cases and climatic factors during monsoon months over a 12-year period are tested, mean relative humidity and lag mean rainfall (i.e., rainfall at one-month lag) were found to be significant. The model explains 93% of the variability in dengue cases while considering climatic factors. As noted earlier, Karim et al.'s [32] study in Dhaka confirmed that rainfall, maximum temperature and relative humidity can explain 61% of the variability in reported dengue cases at a two-month lag. The study also found that the monsoon season, with a peak in August, was highly predictive of most reported dengue cases in Dhaka.

It is evident from both Karim et al.'s [32] and our findings that out of all climatic factors, rainfall has the most influence on dengue incidence in Dhaka. Arcari et al. [54] in Indonesia and Johansson et al. [55] in Puerto Rico also observed a strong positive relationship between rainfall and dengue incidence. Nonetheless, several investigations in Asia and Latin America concluded that the relationship between precipitation and dengue incidence may not be linear, as excess rainfall can negatively impact vector breeding [14,15]. Similar effects were predicted in Dhaka, where continuous heavy rainfall has the potential to flood *Aedes* breeding sites and wash larvae out into fast-flowing rivers, killing them. A significant negative observed association between rainfall and dengue cases, which is reflected in the confidence interval for the co-efficient of rainfall, confirms—albeit indirectly —the negative effects of rainfall on vector abundance and dengue cases. The confidence interval for the coefficients of temperature and relative humidity were positively and significantly associated with dengue cases in Dhaka.

Both short (<5 weeks) and long lag times between climatic factors and increasing dengue incidence have previously been confirmed by studies in Bangladesh [31,32,56] and in Sri Lanka. In Dhaka, Hashizume et al. [31] explain that dengue incidence rarely rises immediately following heavy rainfall. The lag time between rainfall and emerging dengue cases, however, can vary due to several factors. The typical short (<5 week) lag results directly from the life cycle of *Aedes* mosquitos. Heavy rainfall may leave stagnant pools of water on the ground or in objects such as discarded tires, which are ideal habitats for mosquito breeding. The lag time of one month that we found in the present study reflects a relatively more straightforward model of causation—water is required for breeding. Longer models, however, require a different explanation, that is, longer lag times indicate that weather can occur in cycles, and these longer lag times are an artefact solely of the relationship between weather events, rather than between weather events and the thriving of vectors.

Deviations from this standard lag time, however, can emerge from variation in geographic location, elevation, humidity, temperature, infection rate of mosquitoes and other environmental factors. For instance, while in Bangladesh we mostly observed a short lag time (<5 weeks), in our case it was one month, in Sri Lanka, Ehelepola et al. [57] observed an average 5–7-week lag, positively correlated with rainfall, temperature, humidity and hours of sunshine (but negatively correlated with wind). Based on an investigation of dengue incidence in Hanoi, Vietnam, Do et al. [58] distinguished between lag times caused by temperature and rainfall (8–10 weeks lag) and by relative humidity (18 weeks).

The final objective of our study was to determine the pattern of seasonality and climate anomalies and their effects on DF/SDF cases. To this end, we undertook a longitudinal study covering a 13-year period (2001–2013) and assessing temperature, rainfall and relative humidity. Considering significant relationships of dengue incidence with climatic factors, longitudinal studies in terms of seasonality in dengue cases and their variation inter-annually has only recently received wider attention [9,59,60]. In our study, dengue cases exhibited a yearly cyclical pattern, with higher incidence beginning in the monsoon season (June–October), reaching a peak in August and minimum in the pre-monsoon period of January–April in alternate years.

The findings of our longitudinal study revealed that in Dhaka, temperature has a profound effect on vector breeding and DENV spread. The plot of calculated temperature anomalies vs. dengue cases for 2001–2013 exhibits that when the average yearly temperature was 25.9 °C (−0.1 °C anomaly) or 25.8 °C (−0.2 °C anomaly), dengue cases were higher than the average. Contrary to our findings, Beebe et al.'s [61] study in Australia offers an explanation for such effects: higher temperatures in summer months force residents to store water in open containers in their homes, providing ideal breeding sites for *Aedes* mosquitos.

A number of studies worldwide showed strong direct association between temperature and *Aedes* abundance. For example, a study in Brazil revealed that the seasonal pattern of abundance of the *Aedes* mosquito is visible while having the temperature as a parameter at the city level [62]. In Thailand, Chavez et al. [63] observed that *Aedes* mosquito abundance changes considerably with temperature change. Contrary to these findings, an Australian study showed that the development of the immature *Aedes* mosquito was inversely associated with temperature [64]. Conforming with these, the present study in the city of Dhaka, Bangladesh, the mosquito abundance and dengue incidence were observed to be strongly associated with temperature changes. In Dhaka, when annual average temperatures reached 26.1−26.6 °C, dengue incidence was lower than the dengue incidence during "normal" temperatures (26 °C). Overall, our study revealed that dengue incidence rises with annual average temperature up to 26 °C, after which it begins to drop off.

In a study conducted in Bangladesh, Banu et al. [65] found a highly significant association between local climate variables (temperature and rainfall) and dengue incidence. However, when they studied the association between ENSO, IOD and dengue incidence, they observed that the extent of the association was very weak. They also reported that the association between dengue incidence and ENSO or IOD were comparatively stronger after an adjustment for local climate variables, seasonality and trend, when they applied a distributed lag nonlinear model (DLNM). Thus, Banu et al. [64] revealed that stronger effects of local climate variables propel the weak association between ENSO, IOD and dengue incidence, making the local climate variables more important for dengue incidence in Bangladesh.

Similar studies conducted in Puerto Rico, Thailand and Mexico by Johansson et al. [66] also observed a weak association between dengue and ENSO. In an earlier study, Hales et al. [67] found in the 14 island nations of the Pacific that dengue incidence was positively associated with ENSO in 10 of these island countries. Other studies in Thailand and Mexico also reported a positive association between dengue and ENSO [68,69]. Though the study of association between ENSO, IOD and dengue incidence were beyond the scope our study, it is worth noting that the effects of local climatic variables, weather conditions, and other factors regulating dengue dynamics indicate conformity with our study which revealed a strong association of the main climatic variables (temperature, rainfall and humidity) with dengue cases in the city of Dhaka, Bangladesh.

There are some limitations to this study. First, we were constrained by missing data as continuous time series data on dengue cases for the study period were not available in the government depository on dengue surveillance. To overcome this limitation and attain the best possible 'close approximation', the missing data were denoted as "missing at random". Following Little and Rubin [41] and Weerasinghe [42], these missing data were imputed by applying the Spline Interpolation Method (see R-package "imputTS") [43,44] and the regression imputation method [45]. Thus, the data limitation would not pose any serious constraint to the generalizability and future study directions. Second, the entomological surveys that collected data on *Aedes* larvae from households in Dhaka did not cover the same areas throughout the study period. We therefore relied on adjusted data for specific areas which were covered repeatedly by all survey periods over a 5-year period. Third, the entomological data did not cover adult *Aedes* mosquitoes, so, we could not estimate the extrinsic incubation period and the infection rate. Fourth, due to the unavailability of time-series data on other variables, we were unable to incorporate other contributing factors such as household water-use, vector control measures, and land-use changes into our models. Further empirical research on longitudinal trends in seasonality and its effects upon dengue vector breeding will assist with improved understanding of dengue disease transformation.

5. Conclusions

Recognizing that only nominal attempts have thus far been made to empirically examine the dynamic causal relationships within the climate–vector–disease nexus, the present study applied an innovative step-by-step approach to determine the causal relationships between climatic factors, dengue vector abundance, and dengue cases/incidence in the city of Dhaka, Bangladesh. Some of our results are similar to those of other studies, confirming that there is a significant correlation between climatic factors and vector abundance [9,13,48,51], and between vector abundance and dengue incidence [9,10,50]. The examination of climate–vector–disease nexus of our study under one conceptual framework is a unique attempt and thus makes a novel contribution to this research domain concerning dengue disease transmission.

The integration of climate and entomological data into climate–vector–disease nexus to produce predictive models has not yet been fully realized [31]. In this regard, the present study provides evidence of a strong relationship amongst climatic factors, vector and dengue via a step-by-step process—revealing their significance in dengue vector abundance and dengue disease occurrence.

We found that an increase in *Aedes* abundance is associated with the rise in temperature, relative humidity, and rainfall during the monsoon months that in turn appear in the subsequent increase in dengue incidence. The relationships of dengue cases with *Stegomyia* indices as well as with the mean relative humidity and the lag mean rainfall were also highly significant. This study thus lays a strong foundation for future work on dengue forecasting and prevention, which may prove vital as the climate continues to change and the range and seasonal dynamics of dengue and other diseases change with it. As Wilder-Smith et al. [70] highlighted, the predictive models of the relationships between climate variables and dengue transmission that we pursued in the present study can assist in developing early warning systems. Based on the existing literature, we primarily focused on associations between climate data and *Aedes* abundance for the monsoon months. Future study should expand such entomological study for the non-monsoon months as well.

Supplementary Materials: The following are available online at https://www.mdpi.com/article/10.3390/atmos12070905/s1, Figure S1: ACF and PACF plots for CI in relation to the climatic parameters. The PACF plot indicating that the residuals in the fitted model were uncorrelated with mean zero and constant variance; Figure S2: ACF and PACF plots for BI in relation to the climatic parameters. The PACF plot indicating that the residuals in the fitted model were uncorrelated with mean zero and constant variance; Figure S3. Autocorrelation and partial autocorrelation plots for the number of dengue cases vs. climate factors by month for the 13-year study period (2001–2013): (a and b) number of dengue cases: (c and d) average temperature; (e and f) average relative humidity; (g and h) average rainfall; Figure S4. Autocorrelation and partial autocorrelation plot of residuals in the city of Dhaka, Bangladesh for the 13-year study period (2001–2013).

Author Contributions: Conceptualization, S.I. and C.E.H.; methodology, S.I., C.E.H. and S.H.; software, S.H. and S.I.; validation, S.I., S.H. and C.E.H.; formal analysis, S.I., S.H., C.E.H. and J.H.; investigation, S.I. and C.E.H.; resources, S.I., C.E.H., S.H. and J.H.; data curation, S.H. and S.I.; writing—original draft preparation, S.I., C.E.H. and S.H.; writing—review and editing, S.I., C.E.H., S.H. and J.H.; visualization, S.I., C.E.H.; supervision, C.E.H. and S.H.; project administration, C.E.H.; funding acquisition, S.I. and C.E.H. All authors have read and agreed to the published version of the manuscript.

Funding: This research was made possible through funding from the International Development Research Centre (IDRC), Ottawa, Canada (Grant # 106040-001 to the second author); University of Manitoba, Winnipeg, Canada; North South University, Dhaka, and Bangladesh, International Center for Diarrheal Disease Research, Bangladesh (icddr,b), Dhaka, Bangladesh. Further financial assistance was received from the Social Science and Humanities Research Council, InSight Grant (Grant # 435-2012-1748), Ottawa, Canada; and from International Development Research Centre (IDRC), Ottawa, Canada, through the IDRC Doctoral Research Awards (IDRA) (107099-050) to the first author.

Institutional Review Board Statement: This research was approved by the Bangladesh Medical Research Council (Bangladesh) and the Joint Faculty Research Ethics Board of the University of Manitoba (Canada). Administrative permission was granted by the Government of Bangladesh to access and use the meteorological data from the Bangladesh Meteorological Department, and entomological and dengue case data from the Directorate of Public Health and Services.

Informed Consent Statement: Not applicable.

Data Availability Statement: The datasets used and/or analyzed during the current study are available from the corresponding author on reasonable request.

Acknowledgments: This research was funded by the International Development Research Centre (IDRC), Canada (grant # 106040-01 to the 2nd author) and (grant # 107099-050 to the 1st author); and the Social Sciences and Humanities Research Council (SSHRC) of Canada (grant # 435-2018-552) to the 2nd author; we thank these organizations for their financial supports. We are grateful to the Study Participants in Dhaka city communities for having confidence and trust in the interviewers and for sharing personal and household information with them. We are also appreciative of the Bangladesh Meteorological Department and Directorate General of Public Health and Services, Government of Bangladesh for providing us the required data and access to other necessary information and Gilles Messier for his assistance in editing the text.

Conflicts of Interest: The authors declare that they have no competing interest. The funders had no role in study design, data collection and analysis, decision to publish, or preparation of the manuscript.

References

1. Halstead, S.B. Dengue. *Lancet* **2007**, *370*, 1644–1652. [CrossRef]
2. Gubler, D.J. The global pandemic of dengue/dengue haemorrhagic fever: Current status and prospects for the future. *Ann. Acad. Med. Singap.* **1998**, *27*, 227–234. [PubMed]
3. Rodhain, F. Les insectes ne connaissent pas nos frontières. *Médecine Mal. Infect.* **1996**, *26* (Suppl. 3), 408–414. [CrossRef]
4. WHO (World Health Organization). Dengue and Severe Dengue. 2021. Available online: https://www.who.int/news-room/fact-sheets/detail/dengue-and-severe-dengue (accessed on 6 July 2021).
5. Githeko, A.K.; Lindsay, S.W.; Confalonieri, U.E.; Patz, J.A. Climate change and vector-borne diseases: A regional analysis. *Bull. World Health Organ.* **2000**, *78*, 1136.
6. Lorenz, C.; Azevedo, T.S.; Virginio, F.; Aguiar, B.S.; Chiaravalloti-Neto, F.; Suesdek, L. Impact of environmental factors on neglected emerging arboviral diseases. *PLoS Negl. Trop. Dis.* **2017**, *11*, e0005959. [CrossRef]

7. Mcmichael, A.J.; Woodruff, R.E.; Hales, S. Climate change and human health: Present and future risks. *Lancet* **2006**, *367*, 859–869. [CrossRef]

8. Pham, H.V.; Doan, H.T.M.; Phan, T.T.T.; Minh, N.N.T. Ecological factors associated with dengue fever in a central highlands Province, Vietnam. *BMC Infect. Dis.* **2011**, *11*, 172. [CrossRef]

9. Tseng, W.-C.; Chen, C.-C.; Chang, C.-C.; Chu, Y.-H. Estimating the economic impacts of climate change on infectious diseases: A case study on dengue fever in Taiwan. *Clim. Chang.* **2009**, *92*, 123–140. [CrossRef]

10. Stewart Ibarra, A.M.; Ryan, S.J.; Beltrán, E.; Mejía, R.; Silva, M.; Muñoz, Á. Dengue vector dynamics (*Aedes aegypti*) influenced by climate and social factors in Ecuador: Implications for targeted control. *PLoS ONE* **2013**, *8*, e78263. [CrossRef]

11. Depradine, C.A.; Lovell, E.H. Climatological variables and the incidence of dengue disease in Barbados. *Int. J. Environ. Health Res.* **2004**, *14*, 429–441. [CrossRef]

12. Colón-González, F.J.; Fezzi, C.; Lake, I.R.; Hunter, P.R. The effects of weather and climate change on dengue. *PLoS Negl. Trop. Dis.* **2013**, *7*, e2503. [CrossRef]

13. Naish, S.; Dale, P.; Mackenzie, J.S.; McBride, J.; Mengersen, K.; Tong, S. Climate change and dengue: A critical and systematic review of quantitative modelling approaches. *BMC Infect. Dis.* **2014**, *14*, 167. [CrossRef]

14. Halide, H.; Ridd, P. A predictive model for dengue hemorrhagic fever epidemics. *Int. J. Environ. Health Res.* **2008**, *18*, 253–265. [CrossRef]

15. Wu, P.-C.; Lay, J.-G.; Guo, H.-R.; Lin, C.-Y.; Lung, S.-C.; Su, H.-J. Higher temperature and urbanization affect the spatial patterns of dengue fever transmission in subtropical Taiwan. *Sci. Total Environ.* **2009**, *407*, 2224–2233. [CrossRef] [PubMed]

16. Islam, S.; Haque, C.E.; Hossain, S.; Rochon, K. Role of container type, behavioural, and ecological factors in *Aedes* pupal production in Dhaka, Bangladesh: An application of zero-inflated negative binomial model. *Acta Trop.* **2019**, *193*, 50–59. [CrossRef] [PubMed]

17. Kuno, G. Review of the factors modulating dengue transmission. *Epidemiol. Rev.* **1995**, *17*, 321–335. [CrossRef]

18. Wu, J.-Y.; Lun, Z.-R.; James, A.A.; Chen, X.-G. Review: Dengue fever in mainland China. *Am. J. Trop. Med. Hyg.* **2010**, *83*, 664–671. [CrossRef] [PubMed]

19. Yu, H.-L.; Yang, S.-J.; Yen, H.-J.; Christakos, G. A spatio-temporal climate-based model of early dengue fever warning in southern Taiwan. *Stoch Environ. Res. Risk Assess* **2011**, *25*, 485–494. [CrossRef]

20. Shahid, S.; Harun, S.B.; Katimon, A. Changes in diurnal temperature range in Bangladesh during the time period 1961–2008. *Atmos. Res.* **2012**, *118*, 260–270. [CrossRef]

21. Yunus, E.B. Dengue outbreak 2000: The emerged issues. *Bangladesh Med. J.* **2000**, *33*, 46–47.

22. Hossain, M.A.; Khatun, M.; Arjumand, F.; Nisalak, A.; Breiman, R.F. Serologic evidence of dengue infection before onset of epidemic, Bangladesh. *Emerg. Infect. Dis.* **2003**, *9*, 1411–1414. [CrossRef]

23. Yunus, E.B.; Bangali, A.M.; Mahmood, M.A.H.; Rahman, M.M.; Chowdhury, A.R.; Talukder, K.R. Dengue outbreak 2000 in Bangladesh: From speculation to reality and exercises. *Dengue Bull.* **2001**, *25*, 15–20.

24. Sharmin, S.; Glass, K.; Viennet, E.; Harley, D. Interaction of mean temperature and daily fluctuation influences dengue incidence in Dhaka, Bangladesh. *PLoS Negl. Trop. Dis.* **2015**, *9*, e0003901. [CrossRef]

25. Rahman, M.; Rahman, K.; Siddque, A.K.; Shoma, S.; Kamal, A.H.; Ali, K.S.; Nisaluk, A.; Breiman, R.F. First outbreak of dengue hemorrhagic fever, Bangladesh. *Emerg. Infect. Dis.* **2002**, *8*, 738–740. [CrossRef] [PubMed]

26. Morales, I.; Salje, H.; Saha, S.; Gurley, E.S. Seasonal distribution and climatic correlates of dengue disease in Dhaka, Bangladesh. *Am. J. Trop. Med. Hyg.* **2016**, *94*, 1359–1361. [CrossRef] [PubMed]

27. Dhaka Tribune. DGHS: 673 More Hospitalized for Dengue in 24hrs. 13 September 2019. Available online: https://www.dhakatribune.com/bangladesh/2019/09/13/dghs-673-more-hospitalized-for-dengue-in-24hrs (accessed on 14 September 2019).

28. Shirin, T.; Muraduzzaman, A.K.; Alam, A.N.; Sultana, S.; Siddiqua, M.; Khan, M.H.; Akram, A.; Sharif, A.R.; Hossain, S.; Flora, M.S. Largest dengue outbreak of the decade with high fatality may be due to reemergence of DEN-3 serotype in Dhaka, Bangladesh, necessitating immediate public health attention. *New Microbes New Infect.* **2019**, *29*, 100511. [CrossRef]

29. Dhar-Chowdhury, P.; Paul, K.K.; Haque, C.E.; Hossain, S.; Lindsay, L.R.; Dibernardo, A.; Brooks, W.A.; Drebot, M.A. Dengue seroprevalence, seroconversion and risk factors in Dhaka, Bangladesh. *PLoS Negl. Trop. Dis.* **2017**, *11*, e0005475. [CrossRef]

30. Banu, S.; Hu, W.; Hurst, C.; Guo, Y.; Islam, M.Z.; Tong, S. Space-time clusters of dengue fever in Bangladesh. *Trop. Med. Int. Health* **2012**, *17*, 1086–1091. [CrossRef]

31. Hashizume, M.; Dewan, A.K.; Sunahara, T.; Rahman, M.Z.; Yamamoto, T. Hydroclimatological variability and dengue transmission in Dhaka, Bangladesh: A time-series study. *BMC Infect. Dis.* **2012**, *12*, 98. [CrossRef]

32. Karim, M.N.; Munshi, S.U.; Anwar, N.; Alam, M.S. Climatic factors influencing dengue cases in Dhaka city: A model for dengue prediction. *Indian J. Med. Res.* **2012**, *136*, 32–39.

33. Sharmin, S.; Viennet, E.; Glass, K.; Harley, D. The emergence of dengue in Bangladesh: Epidemiology, challenges and future disease risk. *Trans. R. Soc. Trop. Med. Hyg.* **2015**, *109*, 619–627. [CrossRef]

34. Paul, K.K.; Dhar-Chowdhury, P.; Haque, C.E.; Al-Amin, H.M.; Goswami, D.R.; Kafi, M.A.; Drebot, M.A.; Lindsay, L.R.; Ahsan, G.U.; Brooks, W.A. Risk factors for the presence of dengue vector mosquitoes, and determinants of their prevalence and larval site selection in Dhaka, Bangladesh. *PLoS ONE* **2018**, *13*, e0199457. [CrossRef] [PubMed]

35. Islam, M.Z.; Rutherford, S.; Dung Phung, M.; Uzzaman, N.; Baum, S.; Huda, M.M.; Asaduzzaman, M.; Talukder, M.R.; Chu, C. Correlates of climate variability and dengue fever in two metropolitan cities in Bangladesh. *Cureus* **2018**, *10*, e3398. [CrossRef]

36. Ahmed, T.; Rahman, G.M.S.; Bashar, K.; Shamsuzzaman, M.; Samajpati, S.; Sultana, S.; Hossain, M.I.; Banu, N.N.; Rahman, M.S. Seasonal prevalence of dengue vector mosquitoes in Dhaka City, Bangladesh. *Bangladesh J. Zool.* **2007**, *35*, 205–212.
37. BBS (Bangladesh Bureau of Statistics). *Statistical Yearbook 2016*; BBS, Ministry of Planning: Dhaka, Bangladesh, 2016.
38. UN (United Nations). The World's Cities in 2016: Data Booklet. 2016. Available online: http://www.un.org/en/development/desa/population/publications/pdf/urbanization/the_worlds_cities_in_2016_data_booklet.pdf (accessed on 3 June 2020).
39. BBS (Bangladesh Bureau of Statistics). *Statistical Yearbook 2010*; BBS, Ministry of Planning: Dhaka, Bangladesh, 2010.
40. WMO (World Meteorological Organization). *WMO Guidelines on the Calculation of Climate Normals*, 2017 ed.; WMO-No. 1203; WMO: Geneva, Switzerland, 2017; p. 29.
41. Little, R.J.A.; Rubin, D.B. *Statistical Analysis with Missing Data*, 2nd ed.; John Wiley: New York, NY, USA, 2002.
42. Weerasinghe, D.S.S. A missing values imputation method for time series data: An efficient method to investigate the health effects of sulphur dioxide levels. *Environmetrics* **2010**, *21*, 162–172. [CrossRef]
43. Wolberg, G. *Cubic Spline Interpolation: A Review*; Columbia University: New York, NY, USA, 1988.
44. Moritz SBartz-Beielstein, T. imputeTS: Time Series Missing Value Imputation in R. *R J.* **2017**, *9*, 207–218. [CrossRef]
45. Buuren, S.V. *Flexible Imputation of Missing Data*; Chapman and Hall/CRC: Boca Raton, FL, USA; New York, NY, USA, 2012.
46. R Core Team. *R: A Language and Environment for Statistical Computing*; R Foundation for Statistical Computing: Vienna, Austria, 2019; Available online: http://www.R-project.org (accessed on 20 June 2020).
47. Wood, S.N. *Generalized Additive Models: An Introduction with R*; Chapman and Hall: Boca Raton, FL, USA, 2006.
48. Barrera, R.; Amador, M.; MacKay, A.J. Population dynamics of *Aedes aegypti* and dengue as Influenced by weather and human behavior in San Juan, Puerto Rico. *PLoS Negl. Trop. Dis.* **2011**, *5*, e1378. [CrossRef] [PubMed]
49. Bowman, L.R.; Runge-Ranzinger, S.; McCall, P.J. Assessing the relationship between vector indices and dengue transmission: A systematic review of the evidence. *PLoS Negl. Trop. Dis.* **2014**, *8*, e2848. [CrossRef]
50. Githeko, A.K. Advances in developing a climate-based dengue outbreak models in Dhaka, Bangladesh: Challenges and opportunities. Commentary. *Indian J. Med. Res.* **2012**, *136*, 7–9.
51. Cheng, Q.; Jing, Q.; Spear, R.C.; Marshall, J.M.; Yang, Z.; Gong, P. Climate and the timing of imported cases as determinants of the dengue outbreak in Guangzhou, 2014: Evidence from a mathematical model. *PLoS Negl. Trop. Dis.* **2016**, *10*, e0004417. [CrossRef]
52. Nakhapakorn, K.; Tripathi, N.K. An information value-based analysis of physical and climatic factors affecting dengue fever and dengue haemorrhagic fever incidence. *Int. J. Health Geogr.* **2005**, *4*, 13. [CrossRef] [PubMed]
53. Chang, F.-S.; Tseng, Y.-T.; Hsu, P.-S.; Chen, C.-D.; Lian, I.-B.; Chao, D.-Y. Re-assess vector indices threshold as an early warning tool for predicting dengue epidemic in a dengue non-endemic country. *PLoS Negl. Trop. Dis.* **2015**, *9*, e0004043. [CrossRef] [PubMed]
54. Arcari, P.; Tapper, N.; Pfueller, S. Regional variability in relationships between climate and dengue/SD in Indonesia. *Singap. J. Trop. Geogr.* **2007**, *28*, 251–272. [CrossRef]
55. Johansson, M.A.; Dominici, F.; Glass, G.E. Local and global effects of climate on dengue transmission in Puerto Rico. *PLoS Negl. Trop. Dis.* **2009**, *3*, e382. [CrossRef] [PubMed]
56. Muurlink, T.O.; Stephenson, P.; Islam, M.Z.; Taylor-Robinson, A.W. Long-term predictors of dengue outbreaks in Bangladesh: A data mining approach. *Infect. Dis. Model.* **2018**, *3*, 322–330. [CrossRef]
57. Ehelepola, N.D.B.; Ariyaratne, K.; Buddhadasa, W.M.N.P.; Ratnayake, S.; Wickramasinghe, M. A study of the correlation between dengue and weather in Kandy City, Sri Lanka (2003–2012) and lessons learned. *Infect. Dis. Poverty* **2015**, *4*, 42. [CrossRef]
58. Do, T.T.T.; Martens, P.; Luu, N.H.; Wright, P.; Choisy, M. Climatic-driven seasonality of emerging dengue fever in Hanoi, Vietnam. *BMC Public Health* **2014**, *14*, 1078. [CrossRef]
59. Hii, Y.L.; Zhu, H.; Ng, N.; Ng, L.C.; Rocklov, J. Forecast of dengue incidence using temperature and rainfall. *PLoS Negl. Trop. Dis.* **2012**, *6*, e1908. [CrossRef]
60. Bannister-Tyrrell, M.; Williams, C.; Ritchie, S.A.; Rau, G.; Lindesay, J.; Mercer, G.; Harley, D. Weather-driven variation in dengue activity in Australia examined using a process-based modeling approach. *Am. J. Trop. Med. Hyg.* **2013**, *88*, 65–72. [CrossRef]
61. Beebe, N.W.; Cooper, R.D.; Mottram, P.; Sweeney, A.W. Australia's dengue risk driven by human adaptation to climate change. *PLoS Negl. Trop. Dis.* **2009**, *3*, e429. [CrossRef]
62. Lana, R.M.; Morais, M.M.; Lima, T.F.; Carneiro, T.G.; Stolerman, L.M.; dos Santos, J.P.; Cortés, J.J.; Eiras, Á.E.; Codeço, C.T. Assessment of a trap based *Aedes aegypti* surveillance program using mathematical modeling. *PLoS ONE* **2018**, *13*, e0190673. [CrossRef] [PubMed]
63. Chaves, L.F.; Morrison, A.C.; Kitron, U.D.; Scott, T.W. Nonlinear impacts of climatic variability on the density-dependent regulation of an insect vector of disease. *Glob. Chang. Biol.* **2012**, *18*, 457–468. [CrossRef]
64. Tun-Lin, W.; Burkot, T.R.; Kay, B.H. Effects of temperature and larval diet on development rates and survival of the dengue vector *Aedes aegypti* in north Queensland, Australia. *Med. Vet. Entomol.* **2000**, *14*, 31–37. [CrossRef]
65. Banu, S.; Guo, Y.; Hu, W.; Dale, P.; Mackenzie, J.S.; Mengersen, K.; Tong, S. Impacts of El Niño Southern Oscillation and Indian Ocean Dipole on dengue incidence in Bangladesh. *Sci. Rep.* **2015**, *5*, 16105. [CrossRef] [PubMed]
66. Johansson, M.A.; Cummings, D.A.T.; Glass, G.E. Multiyear climate variability and dengue—El Nino Southern Oscillation, weather, and dengue Incidence in Puerto Rico, Mexico, and Thailand: A longitudinal data analysis. *PLoS Med.* **2009**, *6*, e1000168. [CrossRef]

67. Hales, S.; Weinstein, P.; Souares, Y.; Woodward, A. El Nino and the dynamics of vector-borne disease transmission. *Environ. Health Perspect.* **1999**, *107*, 99–102. [PubMed]
68. Tipayamongkholgul, M.; Fang, C.T.; Klinchan, S.; Liu, C.M.; King, C.C. Effects of the El Nino-Southern Oscillation on dengue epidemics in Thailand, 1996–2005. *BMC Public Health* **2009**, *9*, 422. [CrossRef]
69. Hurtado-Diaz, M.; Riojas-Rodriguez, H.; Gomez-Dantes, H.; Cifuentes, E. Short communication: Impact of climate variability on the incidence of dengue in Mexico. *Trop. Med. Int. Health* **2007**, *12*, 1327–1337. [CrossRef]
70. Wilder-Smith, A.; Renhorn, K.E.; Tissera, H.; Abu Bakar, S.; Alphey, L.; Kittayapong, P.; Lindsay, S.; Logan, J.; Hatz, C.; Reiter, P.; et al. Dengue tools: Innovative tools and strategies for the surveillance and control of dengue. *Glob. Health Action* **2012**, *5*, 17273. [CrossRef]

 *atmosphere*

Article

Spatial Recognition of Regional Maximum Floods in Ungauged Watersheds and Investigations of the Influence of Rainfall

Nam-Won Kim [1], Ki-Hyun Kim [2] and Yong Jung [2,*]

[1] Water Resources Research Division, Water Resources and Environment Research Department, Korea Institute of Construction Technology, Goyang 10228, Korea; nwkim@kict.re.kr
[2] Department of Civil and Environmental Engineering, Wonkwang University, Iksan 54538, Korea; hlkh92@gmail.com
* Correspondence: yong_jung@wku.ac.kr

Abstract: This study primarily aims to develop a method for estimating the range of flood sizes in small and medium ungauged watersheds in local river streams. In practice, several water control projects have insufficient streamflow information. To compensate for the lack of data, the streamflow propagation method (SPM) provides streamflow information for ungauged watersheds. The ranges of flood sizes for ungauged watersheds were generated using a specific flood distribution analysis based on the obtained streamflow data. Furthermore, the influence of rainfall information was analyzed to characterize the patterns of specific flood distributions. Rainfall location, intensity, and duration highly affected the shape of the specific flood distribution. Concentrated rainfall locations affected the patterns of the maximum specific flood distribution. The shape and size of the minimum specific flood distribution were dependent on the rainfall intensity and duration. The Creager envelope curve was used to generate equations for the maximum/minimum specific flood distribution for the study site. The ranges of the specific flood distributions were produced for each watershed size.

Keywords: ranges of flood sizes; specific flood distributions; ungauged watersheds; influence of rainfall characteristics

Citation: Kim, N.-W.; Kim, K.-H.; Jung, Y. Spatial Recognition of Regional Maximum Floods in Ungauged Watersheds and Investigations of the Influence of Rainfall. *Atmosphere* **2021**, *12*, 800. https://doi.org/10.3390/atmos 12070800

Academic Editors: Jong-Suk Kim, Nirajan Dhakal, Changhyun Jun and Taesam Lee

Received: 4 May 2021
Accepted: 16 June 2021
Published: 22 June 2021

Publisher's Note: MDPI stays neutral with regard to jurisdictional claims in published maps and institutional affiliations.

1. Introduction

Streamflow data is fundamentally required for the flood design process in water control projects. Many water control infrastructures are built in small/mid-sized regional watersheds, which require records of maximum floods. However, there are scarce observed streamflow data for small-and mid-sized watersheds. Most streamflow observations are obtained at hydrologically interesting locations or at the end of a large watershed located downstream. Many studies estimate maximum regional floods using observed data through various approaches, including empirical, deterministic, and probabilistic methods [1]. Several empirical studies have used regional extreme flood records for empirical flood estimations and other related analyses (e.g., Kovacs [2]: 30 sites at South Africa; Acreman [3]: UK floods for 200-year records; Furey and Gupta [4,5]: data from Goodwin Creek experimental watersheds; Patnaik et al. [6]: 358 basins from the USGS database). For other approaches, rational formulas and flood frequency analyses in deterministic and probabilistic methods also used observed data.

However, many countries, especially developing countries, have limited records of large floods [7]. To overcome this, streamflow data for ungauged watersheds were generated by regionalization or regression analysis [8]. Regionalization used streamflow data near watersheds that have similar characteristics (e.g., physiographic and meteorological features) by transferring streamflow data to ungauged watersheds [9–11]. The regression method was used to search for relationships between streamflow data and watershed characteristics and generate streamflow data for ungauged watersheds [12–14]. As a modified regionalization method, Kim et al. [15] suggested a robust method (i.e., streamflow propagation method (SPM)) to generate streamflow data for ungauged small/mid-sized watersheds. Unlike other regionalization methods, this study aimed to minimize uncertainties using fixed, physically related parameters, while an initial condition (i.e., initial soil saturation) was optimized in a simple lumped conceptual rainfall-runoff model.

In contrast to limited streamflow information, the generated streamflow data can display the spatial distribution of streamflow data for various sizes of watersheds. In this study, we attempt to obtain the ranges of maximum regional floods to aid in the management of small and mid-sized watersheds (less than ~2500 km^2) for which no streamflow data exist. Additionally, the spatial distribution of streamflow for ungauged watersheds and rainfall characteristics on these distributions are explored to generate a range of streamflow magnitude for various sizes of watersheds. The dynamics of rainfall (e.g., rainfall intensity and duration) have substantial effects on the structure of peak discharge [16]. The remaining sections of this paper are organized as follows: Section 2 provides a methodology of this study and Section 3 explains the concept of the streamflow propagation method (SPM) and the study area with selected storm events. The results of streamflow propagation and rainfall effects on the streamflow distribution are presented in Section 4, and Section 4 concludes the study.

2. Methodology

To know the ranges of flood sizes in small- and mid-sized watersheds requires an extensive amount of streamflow data. However, smaller-sized watersheds have a lack of streamflow information frequently due to an abundant amount of timely and economical expenses for monitoring. Thus, the SPM [15] was adopted to generate streamflow data for ungauged smaller watersheds. Kim et al. [15] suggested the spatial propagation of streamflow data using a lumped conceptual model to generate streamflow data for ungauged watersheds. Spatial propagation is achieved when a rainfall-runoff model is run to match the observed streamflow data. Simulations of a rainfall-runoff model for a large watershed can simultaneously produce streamflow information for smaller watersheds within this large watershed. However, several parameters optimized for the simulation can produce many uncertainties in the generated streamflow data [16]. Therefore, a minimum number of parameters were selected for this propagation process to reduce uncertainties as far as possible. The propagation shares a similar magnitude of error Equation (1) to ungauged watersheds within a large watershed, while a rainfall-runoff model attempts to fit streamflow data. In their study, Kim et al. [15] applied the storage function method [17] as a rainfall-runoff model. The model parameters were separated into two sets: physically based parameters and time-variant (i.e., event-based) initial conditions. The physically based parameters highly related to the characteristics of watersheds are fixed, while initial conditions are optimized to fit the streamflow data. This separation provides errors similar to those of gauged watersheds to ungauged watersheds in the propagation process. The substantial differences between this method and the regular regionalization methods are the separation of parameters and focused time for simulations (i.e., not for the prediction but for the past data generation), as shown in Figure 1.

$$\varepsilon_i = \left(\frac{O_i - S_i}{O_i} \right)^2, \ where \ i = 1, 2, \ldots, n \tag{1}$$

$$\varepsilon_1 \cong \varepsilon_2 \cong \ldots \cong \varepsilon_n$$

where ε_i is the error rate at watershed i, and O_i and S_i are the observed and simulated streamflow values at watershed i, respectively.

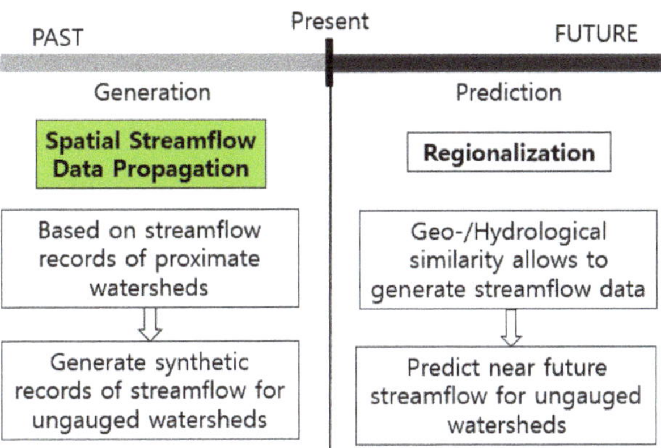

Figure 1. Comparison between the spatial propagation concept and regionalization [15].

After streamflow data generation, streamflow distributions for various sizes of watersheds were observed by using a specific flood distribution analysis, which was usually used to display the streamflow rate for each rainfall event. In this study, we apply this method to display the possible ranges of streamflow amount for the diverse sizes of watersheds. After obtaining the ranges of streamflow, two research questions are addressed: the influence of rainfall on the streamflow distribution and the possible generalization of the range of the streamflow distribution using envelop curve equations to represent (Figure 2). The influence of various rainfall characteristics (e.g., rainfall location, intensity, and duration) was investigated on the effects of the distribution of streamflow for smaller watersheds. For a generalized representation of the range of streamflow distribution (i.e., maximum/minimum specific floods), Creager's envelop curves [18], Kovacs' regional maximum flood [2], and the Korean regional maximum flood equation (KRMF) [19] were tested. The equations of Creager's envelop curves, Kovacs' regional maximum flood, and the KRMF are presented in Equations (2)–(4), respectively.

$$q = 0.503C(0.3861A)^{\left(aA^{b}-1\right)} \tag{2}$$

where q is the specific flood (m^3/s·km^2), C is the Creager coefficient, A is Area (km^2), and a and b are constant variables.

$$Q = 10^6 \left(\frac{A}{10^8}\right)^{1-0.1K} \tag{3}$$

where Q is the streamflow (m^3/s), A is the area (km^2), and K is the coefficient (representing the characteristics of watersheds).

$$q = 11.25\frac{1}{A^{0.25}} \tag{4}$$

where q is the specific flood (m^3/s·km^2), and A is the area (km^2).

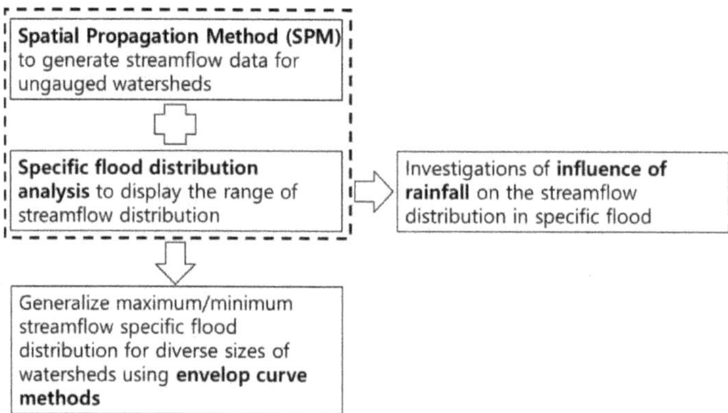

Figure 2. Methodology to represent the range of the streamflow distribution and additional studies related to the specific flood distribution.

3. Study Area and Storm Events

The Han River basin, located near the center of the Korean peninsula, has two major streams: the Northern and Southern Han River streams. The Chungju Dam (CJD) watershed for the study site is located in the upstream part of the Southern Han River stream. The size and length of the CJD watershed are approximately 6648 km^2 and 375 km, respectively. The average width downstream of the CJD is 600 m, with an increased average slope of streamflow. The CJD has a 40% slope for 55.1% of the watershed, and 8.3% of the watershed has a slope of over 80%, meaning that the CJD watershed is located in a mountainous area. Figure 3 shows a streamflow schematic of the CJD watershed. The CJD watershed has two main streams, the left and right sides of the watershed, combined in the middle of the watershed. In this study, the CJD watershed was divided into 35 smaller watersheds with no streamflow information available. To determine the spatial properties of maximum regional floods for a large watershed, spatially distributed streamflow data are necessary. However, most smaller watersheds have limitations in providing streamflow information, as previously mentioned in the Introduction. In the case of the CJD watershed, 12 streamflow gauge stations are insufficient to present the spatial characteristics of the regional maximum flood. Therefore, a propagation process is necessary. We used four different water level gauges (i.e., Chungju Dam (CJD), Youngwol (YW), Youngwol 1 (YW1), and Youngchun (YC)) as validation points, as shown in Figure 3. The CJD inflow was the primary information we wanted to match, while streamflow data propagated to smaller watersheds. The CJD watershed has a significant amount of rainfall during the summer season from June to August, most likely over 60% of the total rainfall. For this study, 18 events with maximum rainfall were selected from 2001 to 2018, as shown in Table 1.

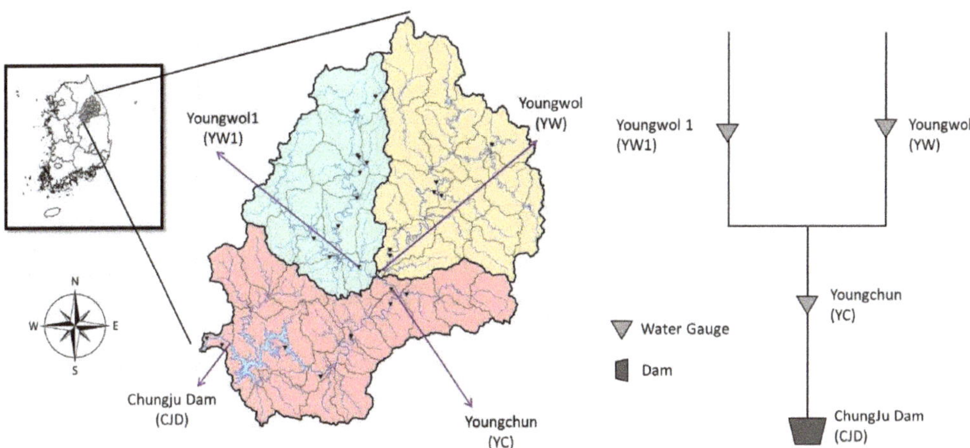

Figure 3. Chungju Dam watershed for study area with two major upstream flows and one combined downstream flow.

Table 1. Selected storm events for CJD watershed.

Event	Start	End	Total Rainfall (mm)	Event	Start	End	Total Rainfall (mm)
1	2001/6/29 0:00	2001/7/4 0:00	101.88	10	2010/9/10 0:00	2010/9/15 0:00	150.33
2	2002/8/6 0:00	2002/8/10 0:00	375.94	11	2011/8/16 0:00	2011/8/19 0:00	145.78
3	2003/7/8 0:00	2003/7/13 0:00	79.86	12	2012/7/5 0:00	2012/7/8 0:00	184.17
4	2004/7/11 0:00	2004/7/15 0:00	136.64	13	2013/7/14 0:00	2013/7/17 0:00	163.16
5	2005/7/1 0:00	2005/7/3 0:00	102.42	14	2014/8/17 0:00	2014/8/23 0:00	106.44
6	2006/7/14 0:00	2006/7/20 0:00	441.94	15	2015/8/24 0:00	2015/8/29 0:00	53.65
7	2007/8/4 0:00	2007/8/7 0:00	126.01	16	2016/7/4 0:00	2016/7/8 0:00	227.03
8	2008/7/24 0:00	2008/7/29 0:00	197.40	17	2017/7/2 0:00	2017/7/6 0:00	190.55
9	2009/7/12 0:00	2009/7/14 0:00	142.90	18	2018/7/1 0:00	2018/7/5 0:00	136.79

4. Results and Discussion

4.1. Generation of Streamflow Data

In the application of a lumped conceptual model (i.e., t storage function method, Kimura 1961) for propagation, there are two parameter sets: physically based parameters (k and p: constant parameters, T_l: lag time) and event-based parameters (f_1: initial runoff ratio, R_{sa}: cumulative saturated rainfall). For the process presented in Figure 4, the cumulative saturated rainfall (R_{sa}) was optimized with an objective function based on the Nash–Sutcliffe efficiency (NSE) at the CJD gauge point while other parameters were fixed. This method delivers a similar magnitude of error for all smaller watersheds. Figure 5 shows the finalized data propagation results using error rates at the validation points. When the soil is fully saturated, the cumulative saturated rainfall is 0, thus suggesting that all rainfall will be fully effective rainfall. The zero cumulative saturated rainfall at events 5, 6, and 9 had previous rainfall events. In terms of error rates, the exit points (i.e., CJD inflow observatory from upstream watersheds) showed higher values at events 3 and 8. The observed values at the CJD increase sharply without any smoother patterns, as shown by other observation points, which means that observations were possibly measured incorrectly. In event 15, the YC and YW have higher error rates because observations at the YC and YW malfunctioned to produce flat values from the beginning of the event (e.g., 34.22 m^3/s at the YC and 34.51 m^3/s at the YW) until a certain point of measurement time, then abruptly increased to the peak streamflow. In addition to these events, Figure 5 shows that the streamflow propagation follows the assumption that similar magnitudes of error rates were distributed, as mentioned in Section 2.

Figure 6 shows the comparisons between the observed and simulated streamflow and peak streamflow at YW, YW1, and YC, respectively. The YC gauge station located below YW1 and YW showed larger streamflow values. The streamflow values at YW were underestimated, while those of YW1 were overestimated. These two contrasting results compensate each other, and the simulated results at the YC show closer streamflow values to the observations because the YC is located under the YW and YW1 gauge stations. The possible reason for contrasting streamflow estimations is the various times of concentration related to the slopes of directions of the YW1 and YW. The fixed parameters are not adjustable based on the status of smaller watersheds in the SPM. The NSEs for peak streamflow at YW1, YW, and YC were 0.94, 0.88, and 0.89, respectively.

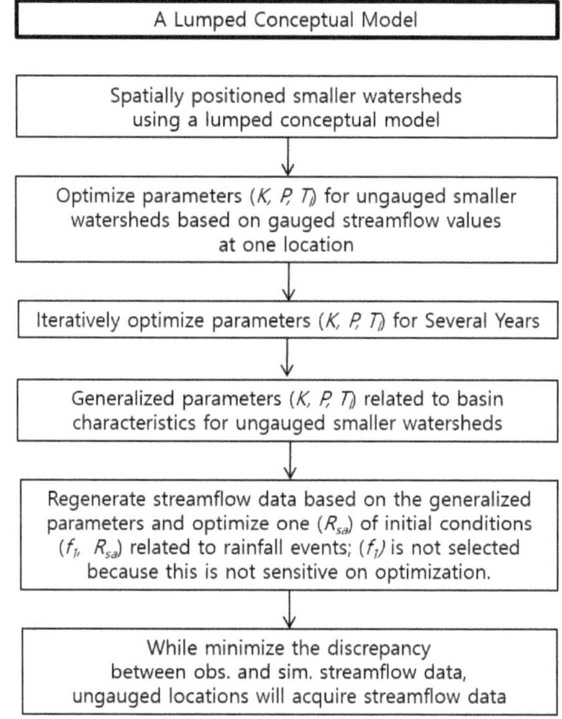

Figure 4. Spatial propagation procedure [15].

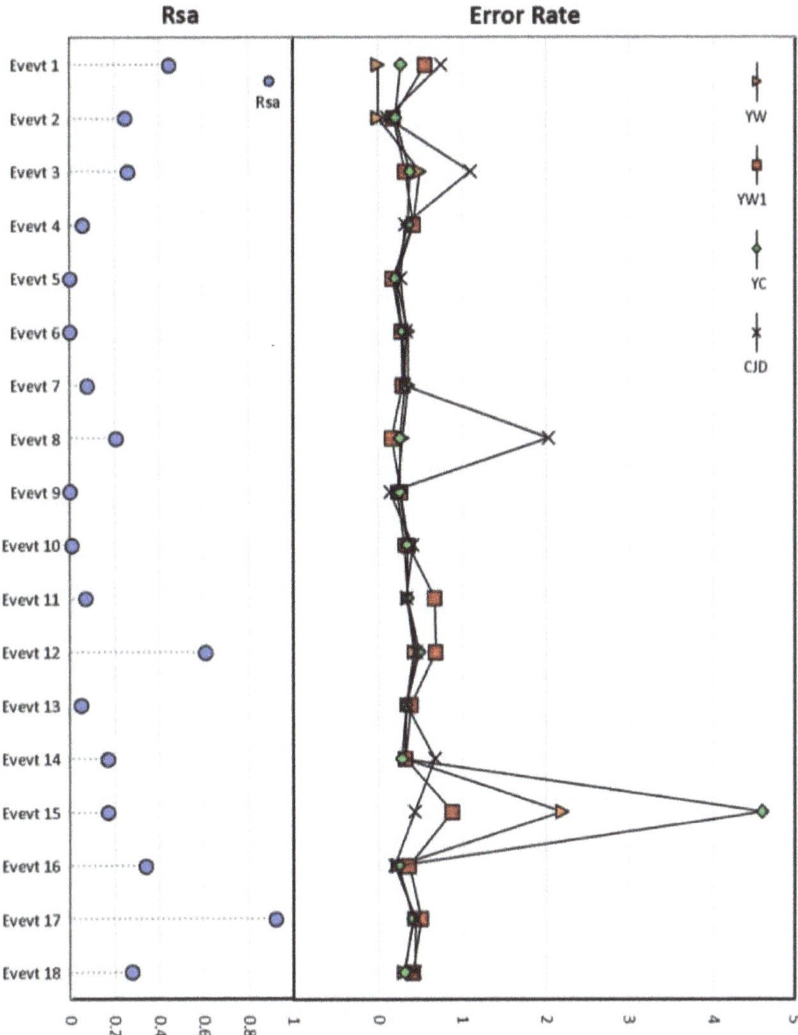

Figure 5. Cumulative saturated rainfall (R_{sa}) and error rates of streamflow data propagation at each gauge station (YW: Youngwol, YW1: Youngwol 1, YC: Youngchun, and CJD: Chungju Dam). Events' details displayed in Table 1.

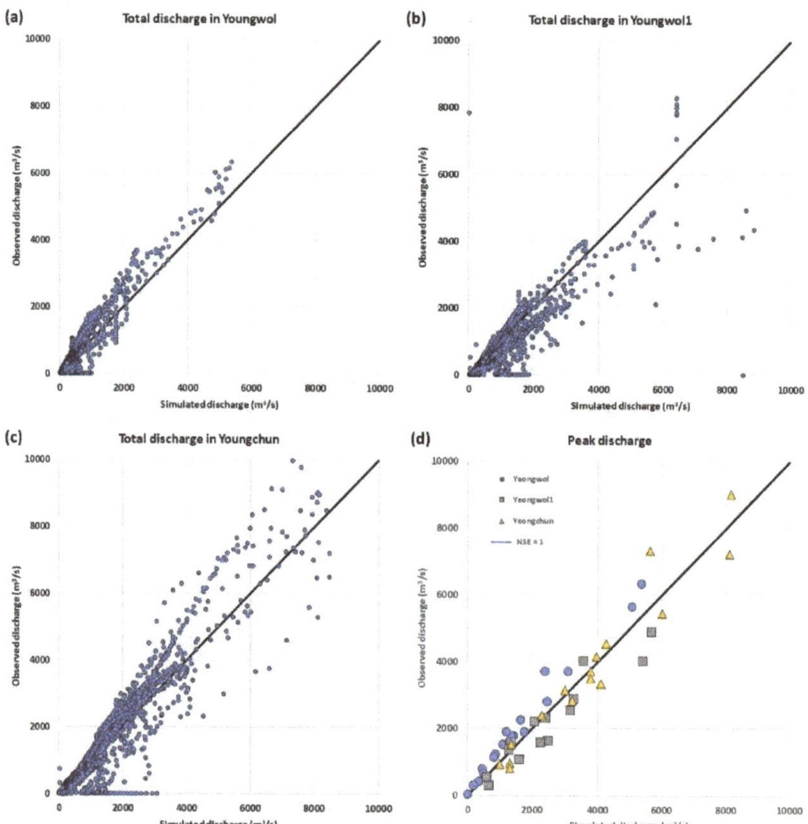

Figure 6. Comparison between observed and simulated peak streamflow values at propagated points. (**a–c**) all discharge comparisons at YW, YW1, and YC, (**d**) peak discharge comparisons.

4.2. Effects of Rainfall on Specific Floods

4.2.1. Rainfall Locations

In this study, a specific flood diagram was adopted to show the rainfall intensities on similarly sized watersheds and display the spatial patterns of streamflow based on the diverse sizes of watersheds. With only observed streamflow data, the patterns of specific floods are barely distinguishable; however, propagated streamflow data may enhance the type of spatial streamflow patterns. The patterns of the spatial distributions of streamflow are validated in Figures 7–9. The spatial distribution of streamflow is highly dependent on rainfall distribution [7]. When rainfall is concentrated on upstream watersheds (i.e., watershed 21, or watersheds 1 and 2, which are highlighted in Figure 7 with watershed numbers) with 20 mm/h (other watersheds with 10 mm/h rainfall) for 24 h, the patterns of the specific flood have power–law profiles, as shown in Figure 7a,b, between specific flood and watershed areas; this has been confirmed by the empirical results of existing research [20–22]. Small watersheds located in the same stream lines displayed as red dots in Figure 7a,b form the power–law pattern in a specific flood. In addition to the power–law patterns, equally distributed rainfall (10 mm/h) has a linearly distributed specific flood below the power–law patterns. In the case of concentrated rainfall downstream (i.e., watershed 32) as shown in Figure 7c, there are no substantial effects for the power–law patterns because the concentrated rainfall location is on downstream watersheds.

Figure 7d displays a sample of specific floods based on the observed rainfall data (event 2013). This specific flood pattern shows the horn shape of the specific flood distribution with a power-law, which means that rainfall was concentrated at upstream watersheds and well distributed with diverse sizes of rainfall.

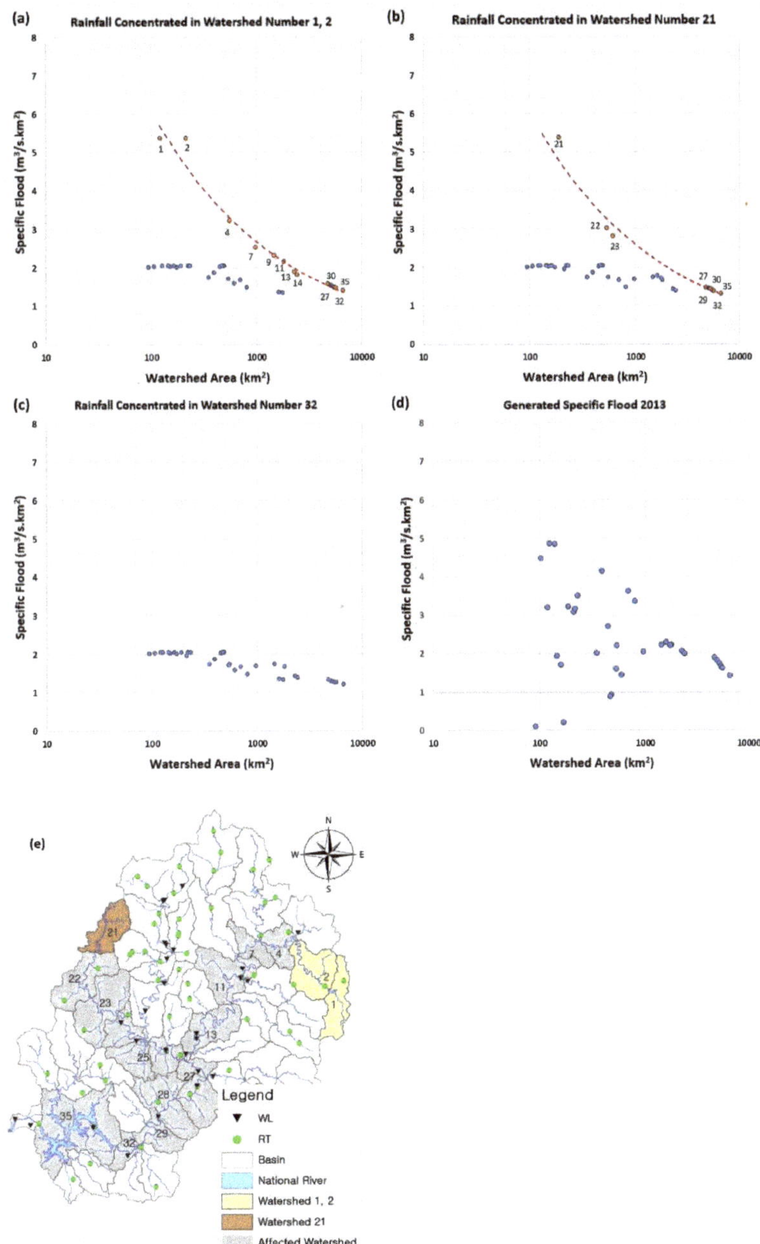

Figure 7. Power–law patterns based on the rainfall concentrated locations (**a**) upstream watershed 21, (**b**) upstream watersheds 1 and 2, (**c**) downstream watershed 32, (**d**) rainfall patterns for an event in 2013), (**e**) concentrated rainfall at watershed 21 (20 mm/hr) or watersheds 1 and 2 (20 mm/hr), with consistent rainfall at other watersheds (10 mm/hr).

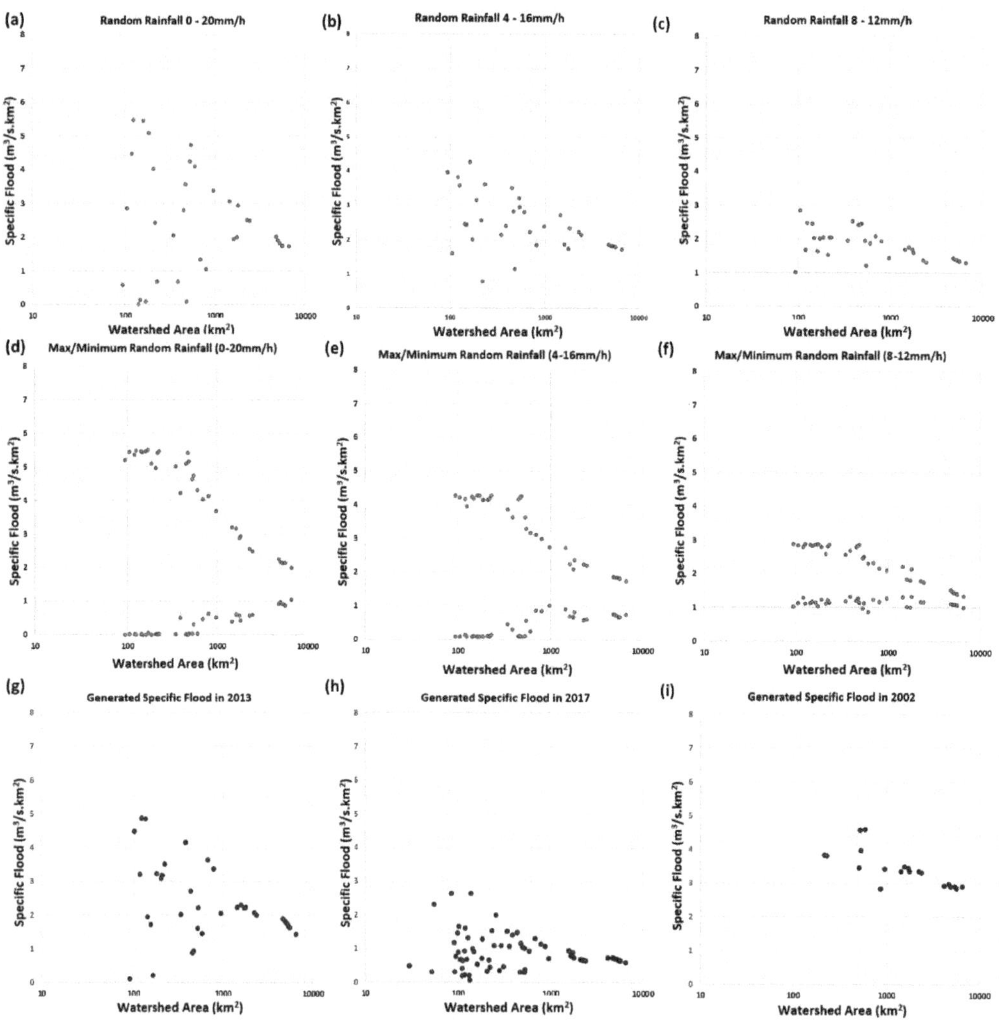

Figure 8. Specific flood based on the different size of rainfall distribution ((**a–c**) Simulation cases, (**d–f**) maximum and minimum size collection of specific floods for 30 trials, (**g–i**) real cases).

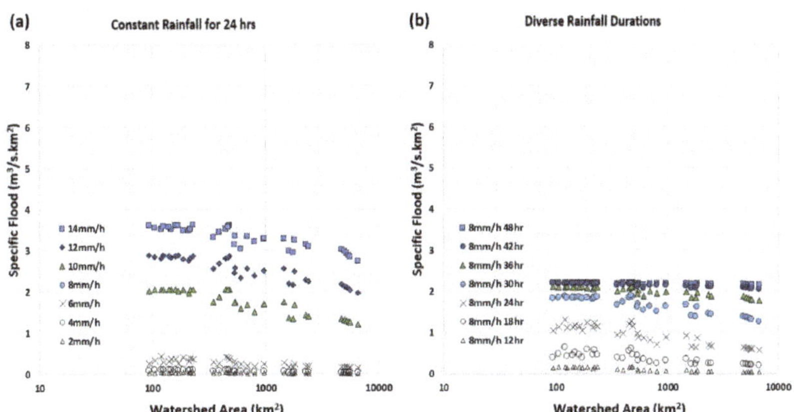

Figure 9. Minimum specific flood dependent on rainfall intensity and rainfall duration.

4.2.2. Rainfall Intensity and Duration

The effects of different sizes of rainfall on the specific flood distribution were investigated in a large watershed. Based on the given ranges of rainfall sizes (i.e., 0–20 mm/h, 4–16 mm/h, and 8–12 mm/h) for 24 h, rainfall sizes were randomly selected and distributed at 35 smaller watersheds. Using randomly selected rainfall values, the Thiessen polygon method provides average rainfall for the CJD watershed. A total of 30 trials were performed within a specified rainfall range in a random rainfall-size-selection manner to ensure specific flood distribution patterns. For comparison purposes, the coefficient of variance (CV) of the rainfall size distribution was applied for each event. Figure 8a–c displays the specific flood of one of the 30 trials, and Figure 8d–f shows the collection of the maximum and minimum specific flood of 30 trials with various ranges of rainfall sizes to show the possible ranges of specific flood for each rainfall range case. Additionally, Figure 8g–i presents the actual events of a specific flood. In the case of 0–20 mm/h (average CV: 59.52), the specific flood distribution displays a widely spread fan shape. If the range of rainfall size is decreased to 8–12 mm/h (average CV: 13.95), the fan shape range is decreased. These results follow the peak discharge distribution trend depending on the spatial CV of total rainfall [6,7]. In Figure 8g–i, the 2013, 2017, and 2002 flood events close to the selected range of rainfall distribution are displayed. The event in 2013 had a wider range of rainfall (CV: 48) and well-distributed rainfall in the CJD watershed. Following the sample patterns of specific floods in Figure 8b,c, the events in 2010 and 2016 have smaller rainfall distribution ranges with smaller CVs of 24.36 and 8.21. From Figure 8, the variances of the specific flood distribution dependent on the variability of rainfall were substantially decreased when the watershed area was larger than 1100 km², which is similar to the results of Mandapaka et al. [23], i.e., the basin responses to the variability of rainfall are dampened over 1000 km² of watersheds. Based on these results, we can determine the range of rainfall distribution from the diverse sizes of the fan shape in the specific flood distribution.

Another reason for the various sizes of specific floods is diverse rainfall durations with various rainfall intensities. To determine the minimum values of specific floods by providing the range of floods, 24 h rainfall with diverse sizes of rainfall intensity was applied. When diverse sizes of rainfall intensity were applied, each simulation was initially designated with the same size of rainfall intensity (i.e., 2, 4, 6, 8, 12, and 14 mm/h) for 35 smaller watersheds. The selected sizes of rainfall intensities were generally found in actual rainfall observations. Figure 9a shows that the constantly applied rainfall intensity for 24 h generates a horizontally distributed specific flood with some decreased values for larger watershed sizes. The smaller watersheds have a shorter time of concentration, providing a more horizontally straight line. However, larger watersheds require more time to reach the time of concentration. The time of concentration for a watershed smaller than 450 km^2 was less than 24 h for the CJD watershed, which is approximately within the range of the results of Abdullah et al. [7]. Their results on the relationship between rainfall duration and peak-specific discharge estimated that maximum discharges were obtained for rainfall durations of 5 to 15 h for medium-sized watersheds (200–300 km^2). In Figure 9b, the same rainfall intensity at 8 mm/h was applied for diverse rainfall durations. As previously mentioned, when the time of rainfall duration reached 40 or 48 h, the larger watersheds also had a horizontally straight specific flood distribution, as shown by the smaller watersheds. The size of the specific floods in Figure 9b decreased substantially after 36 h. Ayalew et al. [16] and Furey and Gupta [4] also showed rainfall duration effects on the peak discharge in a power-law relation ($Q(A) = \alpha A^\theta$, where $Q(A)$ is the peak discharge, α is the intercept, A is the drainage area, and θ is the exponent. They also mentioned that the increasing rates of intercept and exponent after 36 h were also substantially decreased, similar to our results. Additionally, we found that the increased size of watersheds slowly decreased specific floods, which is different from the real cases (e.g., increased specific flood for larger watersheds, as shown in Figure 8d for the minimum specific flood values). In reality, for many cases, smaller watersheds primarily located upstream often have no rainfall while other locations have rainfall, which can give a specific flood close to zero. Contrarily, the larger watersheds will have some specific flood values because at least some rainfall will be included in the larger area. The reason for the different shapes of the minimum specific flood values is that the simulation has rainfall information for all watersheds. The rainfall intensity is the primary factor that increases the magnitude of a specific flood because Figure 9a shows that the size of a specific flood sensitively increases with the rainfall intensity instead of the rainfall duration [7].

4.3. Ranges of Regional Floods

The envelop curve method was used to establish regional relationships between maximum streamflow data and watershed area using the streamflow data generated at ungauged watersheds. We assumed that the study region is hydroclimatically homogenous, and the size of the watershed area is the primary watershed characteristic controlling the streamflow size [24]. Figure 10 displays the representative regional maximum floods for the CJD watershed. The streamflow data generated using the proposed study were compared with various regional maximum flood equations. Figure 10a shows the applications of Creager's envelope curves [18].

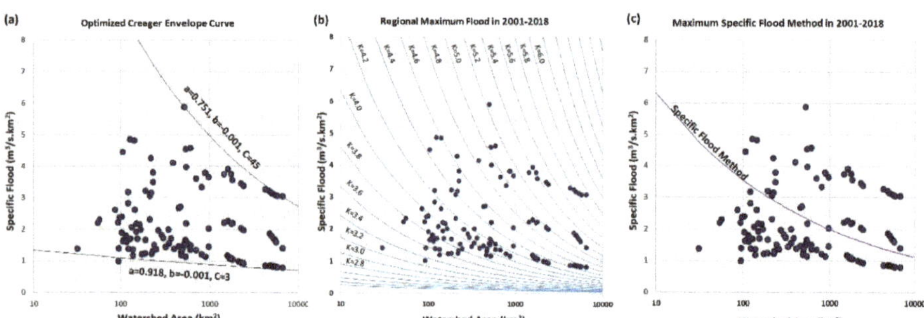

Figure 10. Specific flood for all events with applications of various regional maximum flood equations in 2001–2018. (**a**) Creager's envelop curve, (**b**) Kovacs' regional maximum flood, and (**c**) Korean regional maximum flood).

The upper/lower Creager curve displays the maximum/minimum of the specific flood at each watershed size. The equations of the upper and lower Creager curves use the following values: C = 45, a = 0.751, and b = −0.001; and C = 3, a = 0.918, and b = −0.001, respectively. As shown in Figure 10a, smaller watersheds have a larger variance of a specific flood because these can be sensitively affected by the various rainfall sizes. However, larger watersheds had little influence on the various rainfall sizes. It is evident that any specific size of the watershed can have an explicit range of specific sizes of floods (e.g., 1–5 m^3/s·km^2 for 1000 km^2). Figure 10b shows the Kovacs' regional specific flood equation.

The given specific flood in the CJD watershed can range between K = 3.0 and K = 6.2 in the Kovacs' coefficient to represent the characteristics of watersheds. However, this equation shows the limitation of showing the pattern of the specific flood distribution. Additionally, Figure 10c was plotted using the Korean regional maximum flood equation [19] for a specific flood for the CJD watershed. The specific flood generated for 2001–2009 was not in the range of the maximum floods provided by the KRMF. Therefore, this equation should be modified to be applicable to real cases of specific floods in Korea.

5. Conclusions

The specific flood diagram may have the potential to show the regional streamflow distribution based on the various sizes of floods and watersheds. However, the lack of streamflow data shows the limitation of providing adequate information on streamflow distribution. Thus, the streamflow propagation concept was adopted to provide more spatially distributed streamflow data for ungauged watersheds.

In the process of building a specific flood diagram using the streamflow propagation concept, rainfall information highly influences the distribution patterns of specific floods. The power–law relations were observed to be dependent on the concentrated location of rainfall. If the rainfall is concentrated upstream of the watersheds, the stream located in the same streamline shows a strong power–law profile. A large coefficient variance (CV: 59.52, CV: 13.95) of rainfall intensity ranges reveals a wide fan shape of the specific flood diagram for the smaller watersheds. Rainfall duration and selected size of rainfall were tested for the effects on the minimum values of the specific flood to provide the range of specific floods for diverse watershed sizes. A watershed larger than 450 km^2 requires more time to provide more horizontally distributed minimum values of specific floods. Creager's envelope curve was applied to obtain the maximum and minimum specific flood at various watershed sizes. A watershed size of fewer than 1000 km^2 has a specific range of 1–5 (m^3/s)(km^2). The range of specific floods can be generated for ungauged watersheds of various sizes. Additionally, Kovacs' curve is not suitable for this CJD watershed, and the KRMF must be modified to generate maximum streamflow data for ungauged watersheds. For further study, more rainfall or streamflow data can generalize the range of specific floods for the specified watersheds.

Author Contributions: N.-W.K. and Y.J. conceived and designed the study. K.-H.K. performed calculations. K.-H.K. and Y.J. prepared the original draft. Y.J. supervided the study and revised the manuscript. All authors have read and agreed to the published version of the manuscript.

Funding: This research received no external funding.

Acknowledgments: This study was supported by Wonkwang University in 2021. We are truly grateful for the model support (COSFIM) by K-water in Korea.

Conflicts of Interest: The authors declare no conflict of interest.

References

1. Pegram, G.; Parak, M. A review of the regional maximum flood and rational formula using geomorphological information and observed floods. *Water SA* **2004**, *30*, 377–392. [CrossRef]
2. Kovacs, Z. *Regional Maximum Flood Peaks in Southern Africa*; Technical Report No. 137; Department of Water Affairs: Pretoria, South Africa, 1988.
3. Acreman, M.C. Extreme historical UK Floods and maximum flood estimation. *Water Environ. J.* **1989**, *3*, 404–412. [CrossRef]
4. Furey, P.R.; Gupta, V.K. Effects of excess rainfall on the temporal variability of observed peak-discharge power laws. *Adv. Water Resour.* **2005**, *28*, 1240–1253. [CrossRef]
5. Furey, P.R.; Gupta, V.K. Diagnosing peak-discharge power laws observed in rainfall–runoff events in Goodwin Creek experimental watershed. *Adv. Water Resour.* **2007**, *30*, 2387–2399. [CrossRef]

6. Patnaik, S.; Biswal, B.; Kumar, D.N.; Sivakumar, B. Effect of catchment characteristics on the relationship between past discharge and the power law recession coefficient. *J. Hydrol.* **2015**, *528*, 321–328. [CrossRef]
7. Abdullah, J.; Muhammad, N.S.; Muhammad, S.A.; Julien, P.Y. Envelope curves for the specific discharge of extreme floods in Malaysia. *J. Hydro-Environ. Res.* **2019**, *25*, 1–11. [CrossRef]
8. Sandrock, G.; Viraraghavan, T.; Fuller, G.A. Estimation of peak flows for natural ungauged watersheds in southern Saskatchewan. *Can. Water Resour. J.* **1992**, *17*, 21–31. [CrossRef]
9. McDonnell, J.J.; Beven, K. Debates-The future of hydrological sciences: A (common) path forward? A call to action aimed at understanding velocities, celerities and residence time distributions of the headwater hydrograph. *Water Resour. Res.* **2014**, *50*, 5342–5350. [CrossRef]
10. Merz, R.; Bloschl, G. Regionalisation of catchment model parameters. *J. Hydrol.* **2004**, *287*, 95–123. [CrossRef]
11. Oudin, L.; Andréassian, V.C.; Perrin, C.; Michel, C.; Le Moine, N. Spatial proximity, physical similarity, regression and ungauged catchments: A comparison of regionalization approaches based on 913 French catchments. *Water Resour. Res.* **2008**, *44*, W03413. [CrossRef]
12. Heřmanovský, M.; Havlíček, V.; Hanel, M.; Pech, P. Regionalization of runoff models derived by genetic programming. *J. Hydrol.* **2017**, *547*, 544–556. [CrossRef]
13. Klotz, D.; Herrnegger, M.; Schulz, K. Symbolic Regression for the Estimation of Transfer Functions of Hydrological Models. *Water Resour. Res.* **2017**, *53*, 9402–9423. [CrossRef]
14. Wagener, T.; Wheater, H.S. Parameter estimation and regionalization for continuous rainfall-runoff models including uncertainty. *J. Hydrol.* **2005**, *320*, 132–154. [CrossRef]
15. Kim, N.W.; Jung, Y.; Lee, J.E. Spatial propagation of streamflow data in ungauged watersheds using a lumped conceptual model. *J. Water Clim. Chang.* **2019**, *10*, 89–101. [CrossRef]
16. Ayalew, T.B.; Krajewski, W.F.; Mantilla, R.; Small, S.J. Exploring the effects of hillslope-channel link dynamics and excess rainfall properties on the scaling structure of peak-discharge. *Adv. Water Resour.* **2014**, *64*, 9–20. [CrossRef]
17. Kimura, T. *The Flood Runoff Analysis Method by the Storage Function Model*; The Public Works of Research Institute, Ministry of Construction: Tokyo, Japan, 1961.
18. Creager, W.P.; Justin, J.D.; Hinds, J. *Engineering for Dams*; General Design; John Wiley: New York, NY, USA, 1945; Volume 1.
19. Ministry of Construction. *Report of Water Resource Management in Korea: Guideline of Flood Design Estimation*; Ministry of Construction: Seoul, Korea, 1993.
20. Gupta, V.K.; Mantilla, R.; Troutman, B.M.; Dawdy, D.; Krajewski, W.F. Generalizing a nonlinear geophysical flood theory to medium-sized river networks. *Geophys. Res. Lett.* **2010**, *37*, 11402. [CrossRef]
21. Lima, C.H.R.; Lall, U. Spatial scaling in a changing climate: A hierarchical Bayesian model for non-stationary multi-site annual maximum and monthly streamflow. *J. Hydrol.* **2010**, *383*, 307–318. [CrossRef]
22. Smith, J.A.; Baeck, M.L.; Villarini, G.; Krajewski, W.F. The hydrology and hydrometeorology of flooding in the Delaware River basin. *J. Hydrometeorol.* **2010**, *11*, 841–859. [CrossRef]
23. Mandapaka, P.V.; Krajewski, W.F.; Mantilla, R.; Gupta, V.K. Dissecting the effect of rainfall variability on the statistical structure of peak flows. *Adv. Water Resour.* **2009**, *32*, 1508–1525. [CrossRef]
24. Bayazit, M.; Onoz, B. Envelope curve for maximum floods in Turkey. *Digest* **2004**, 927–931.

 *atmosphere*

Article

Decision-Tree-Based Classification of Lifetime Maximum Intensity of Tropical Cyclones in the Tropical Western North Pacific

Sung-Hun Kim [1], Il-Ju Moon [2,*], Seong-Hee Won [3], Hyoun-Woo Kang [1] and Sok Kuh Kang [1]

[1] Korea Institute of Ocean Science and Technology, Busan 49111, Korea; sh.kim@kiost.ac.kr (S.-H.K.); hwkang@kiost.ac.kr (H.-W.K.); skkang@kiost.ac.kr (S.K.K.)
[2] Typhoon Research Center, Jeju National University, Jeju City 63241, Korea
[3] National Typhoon Center, Korea Meteorological Administration, Jeju City 63614, Korea; shwon11@korea.kr
* Correspondence: ijmoon@jejunu.ac.kr

Abstract: The National Typhoon Center of the Korea Meteorological Administration developed a statistical–dynamical typhoon intensity prediction model for the western North Pacific, the CSTIPS-DAT, using a track-pattern clustering technique. The model led to significant improvements in the prediction of the intensity of tropical cyclones (TCs). However, relatively large errors have been found in a cluster located in the tropical western North Pacific (TWNP), mainly because of the large predictand variance. In this study, a decision-tree algorithm was employed to reduce the predictand variance for TCs in the TWNP. The tree predicts the likelihood of a TC reaching a maximum lifetime intensity greater than 70 knots at its genesis. The developed four rules suggest that the pre-existing ocean thermal structures along the track and the latitude of a TC's position play significant roles in the determination of its intensity. The developed decision-tree classification exhibited 90.0% and 80.5% accuracy in the training and test periods, respectively. These results suggest that intensity prediction with the CSTIPS-DAT can be further improved by developing independent statistical models for TC groups classified by the present algorithm.

Keywords: tropical cyclone; depth-averaged temperature; decision tree; lifetime maximum intensity

Citation: Kim, S.-H.; Moon, I.-J.; Won, S.-H.; Kang, H.-W.; Kang, S.K. Decision-Tree-Based Classification of Lifetime Maximum Intensity of Tropical Cyclones in the Tropical Western North Pacific. *Atmosphere* **2021**, *12*, 802. https://doi.org/10.3390/atmos12070802

Academic Editor: Corene Matyas

Received: 17 May 2021
Accepted: 19 June 2021
Published: 22 June 2021

Publisher's Note: MDPI stays neutral with regard to jurisdictional claims in published maps and institutional affiliations.

1. Introduction

The accurate prediction of tropical cyclone (TC) intensity is a major task in operational forecasting. Regarding intensity prediction, the capabilities of the widely used traditional statistical approaches have improved considerably more than those of the dynamical models [1]. A new statistical–dynamical model, the CSTIPS-DAT [2], which uses a clustering technique and depth-averaged ocean temperature (DAT)-based predictors, has facilitated significant improvements in intensity prediction in the western North Pacific (WNP). However, the CSTIPS-DAT shows relatively large errors for specific clusters, particularly those with a large predictand variance [2].

The tropical western North Pacific (TWNP) TCs, which belong to Cluster 2 in the CSTIPS-DAT model, spend most of their lifetimes over the tropics, where the environmental factors are favorable for their development (Figure 1a). Therefore, Cluster 2 is characterized by the strongest mean TC intensity in the WNP, and many TCs in the said cluster are distinguished by noticeable intensification. However, a considerable number of TCs in the said cluster still do not intensify even under favorable conditions, which produces a large breadth of intensity distribution (Figure 2a) and a large predictand variance. The distribution of the lifetime maximum intensity (LMI) in the TWNP is bimodal, characterized by a local minimum (at about 70 knots LMI) that separates the two groups between weakly (1st mode) and strongly developing TCs (2nd mode). Because the CSTIPS-DAT is a multiple-linear-regression-based model, the TWNP cluster was trained to fit well with strong TCs with a high-density distribution; thus, major errors can occur in the prediction of weak TCs. For example, the intense Typhoon Phanfone in 2014, with an LMI of 95 kt, was well predicted by the CSTIPS-DAT. However, the relatively weak Typhoon Faxai (2014) was not accurately predicted, mostly because of overestimation (Figure 1b,c). These results suggest that with prior knowledge of the LMI type at the genesis of a TC, intensity prediction in the TWNP could be improved through the development of independent statistical models for each classified group.

The LMI, which is an integrated metric of TC intensification, can be used to present basic TC climatology characteristics [3–5]. Several studies have noted that the global distribution of the LMI is bimodal [6–8]. However, there is no consensus on why this bimodal LMI distribution occurs. Torn and Snyder [9] argued that the bimodality is the result of an artificially low number of Category 3 hurricanes in the Atlantic, and that this may be linked to the low resolution of the Dvorak technique which has been used to estimate their intensity. Soloviev et al. [10] attempted to explain the bimodal distribution of the LMI by using the ratio of surface exchange coefficients as a function of wind speed. They suggested that a local maximum of the ratio is favorable for rapid intensification (RI), and thereby increases the number of TCs in the second high-intensity peak. Lee et al. [8] reported that RI is be a key factor in the bimodality in the LMI distribution of two types of TCs: those that undergo RI during their lifetimes (RI TCs) and those that do not (non-RI TCs). They found that the LMI had a normal distribution with a unimodal peak for each TC type, at approximately 120 kt and 45 kt for RI TCs and non-RI TCs, respectively. The establishment of classification criteria to determine the types of TCs (weakly or strongly developing TCs) in the early developing stages of the TWNP TCs will contribute to a better understanding of the global bimodal LMI distribution.

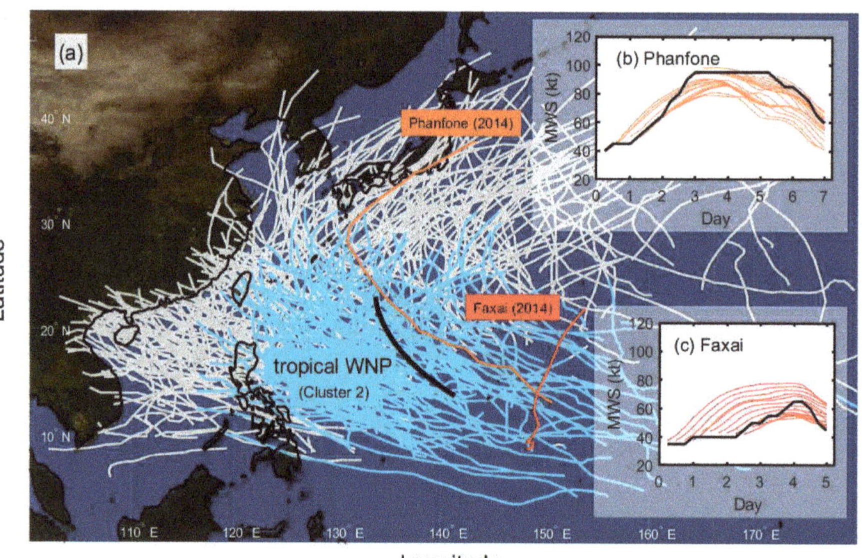

Figure 1. (**a**) All tracks in the tropical western North Pacific (Cluster 2, blue lines), and all tracks in the western North Pacific (gray lines) in 2004–2014. The orange and red lines indicate the tracks of Typhoon Phanfone and Faxai in 2014, respectively. The thick black line is the mean track for Cluster 2 in CSTIPS-DAT. Results of individual intensity predictions from CSTIPS-DAT for Typhoons (**b**) Phanfone and (**c**) Faxai in 2014. A thick black line is an observation (Regional Specialized Meteorological Center best track data), and the colored lines are individual CSTIPS-DAT predictions.

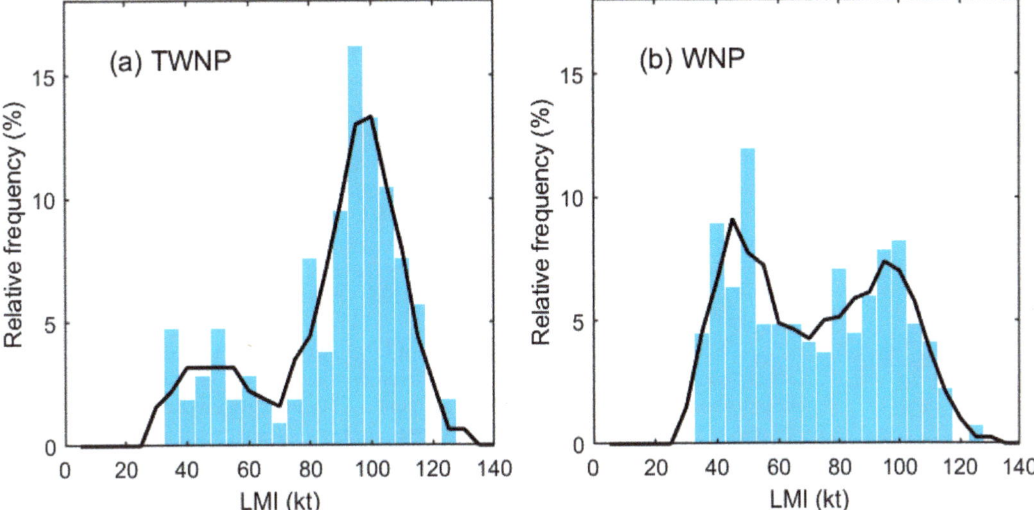

Figure 2. Distribution of lifetime maximum intensity. The relative frequencies presented were calculated on the basis of the 2004–2016 tropical cyclones in (**a**) the tropical western North Pacific (i.e., Cluster 2 in CSTIPS-DAT) and (**b**) the western North Pacific. The blue bars show the raw data binned into 5 kt bins. The black lines present the smoothed relative frequencies with a window width of 15 kt.

The analysis of climate information on TCs has socioeconomic implications and scientific significance because it leads to a better understanding of TC activity and the related mechanisms [11–13]. However, the large volume of varied data on TCs has continued to increase significantly, at a pace that has seemed to outstrip the capabilities of traditional analytical methods [14,15]. The decision tree, as a data-mining technique, is a process of finding useful rules, patterns, and knowledge in large, diverse archived databases to facilitate decision making [16].

Recently, the decision tree, as a useful tool for schematic classification, has been widely employed to investigate the mechanisms of TC development and impact in the WNP [14,17–23] and the North Atlantic [24–27]. Li et al. [17] employed a decision-tree algorithm to investigate the collective contributions to Atlantic hurricanes from sea surface temperature (SST), water vapor, vertical wind shear, and zonal stretching deformation. Zhang et al. [14] applied a decision tree to the binary classification of TCs as intensifying or weakening within 24 h. The decision tree, which used only three variables, exhibited remarkable prediction accuracy: 90.2%. Zhang et al. [18] used a decision tree to investigate the classification of tropical disturbances that did or did not develop into tropical storms in the WNP. The classification accuracies of the developed model were 81.7% for training and 84.6% for validation. Gao et al. [19] used a decision-tree algorithm to develop an RI prediction model that classified intensity changes as RI and non-RI events. They showed that the prestorm ocean coupling potential intensity index, which uses DATs instead of SST to calculate the maximum potential intensity (MPI), improved the RI classification accuracy by approximately 6% during the test period. Park et al. [20] developed a decision-tree-based WNP TC genesis detection algorithm using satellite observation-based predictors. They found that circulation symmetry and intensity were the most critical parameters for characterizing the development of tropical disturbances. Lee et al. [21] developed a scheme for TC formation using machine learning in the WNP and applied it for operational prediction of TC formation. Kim et al. [22] compared the prediction performance of three machine-learning algorithms (decision tree, random forest, support vector machines) and a linear-discriminant-analysis-based model in WNP TC genesis detection. They showed that machine-learning-based models were more capable than conventional linear approaches at detecting TC formation. Yang et al. [24] showed that using the association rule algorithm, the RI prediction performance of the model using only three predictors was better than that of the model consisting of five predictors proposed by Kaplan and DeMaria [28]. Yang [25] performed RI prediction using various classifiers based on the Statistical Hurricane Intensity Prediction Scheme (SHIPS) database. Su et al. [26], using satellite-observation-based storm internal structure and the predictors of the National Hurricane Center probabilistic forecast guidance, developed an RI prediction model for Atlantic hurricanes based on a machine-learning method. Wei and Yang [27] built an artificial intelligence system based on the SHIPS database, significantly improving the RI prediction performance for Atlantic hurricanes. These studies have shown that decision trees are useful for binary classification related to TC genesis and intensification, which further suggests that a decision tree could be a useful tool to split the components in the bimodal distribution of the LMI.

The bimodal distribution results in a large variance of TC intensity, which makes accurate intensity predictions difficult. Therefore, if we can successfully classify the type of LMI at the point when a TC occurs, the statistical TC intensity prediction can be improved by reducing the variance of the predictand. To check such a possibility, this study aimed to build a decision-tree classifier that can predict the intensification type when a target TC occurs. Section 2 describes the dataset and the classification method. In Section 3, the potential predictors are examined, and the classification and model verification results are discussed. A summary and conclusion are provided in Section 4.

2. Data and Methodology

2.1. Data

A decision tree was trained using the 2004–2013 TWNP TCs, which belong to Cluster 2 as classified by the TC track pattern clustering method [2]. Meanwhile, the tree was validated using the 2014–2016 TCs. The TC information was obtained from the Regional Specialized Meteorological Center's best track data. The environmental data were derived from two dynamical models' analysis data. The atmospheric variables were obtained from the National Centers for Environmental Prediction Global Forecast System analysis data, with a 1 × 1 degree of horizontal resolution at 6 h intervals. The oceanic variables were calculated with three-dimensional ocean data derived from the Hybrid Coordinate Ocean Model (HYCOM) + Navy Coupled Ocean Data Assimilation Global Analysis (GLBa0.08) provided by the U.S. Naval Research Laboratory.

2.2. Methodology

2.2.1. Static and Synoptic Potential Predictors

A total of 38 variables were used to build the decision tree, and are listed in Table 1 with their correlations with LMI. The potential variables considered in this study are factors known to be related to TC intensity [2], and are similar to those considered for the development of the CSTIPS-DAT. Four static variables were included: the absolute Julian day number, TC latitude (LAT) and longitude, and TC translation speed. There were 34 synoptic variables: divergence at 200 hPa (D200), the relative vorticity at 500 hPa (RV500) and 850 hPa (RV850), 200 hPa zonal wind (U200) and air temperature (T200), 500–300 hPa layer mean relative humidity (RHHI), 850–700 hPa layer mean relative humidity (RHLO), 200–850 hPa vertical wind shear (SH200), 500–850 hPa vertical wind shear (SH500), ocean heat content (OHC), depth-averaged temperature at various depths (DAT; [29,30]), and DAT-based MPI (DMPI; [31,32]). Lin et al. [31] suggested DMPI using DAT instead of prestorm sea surface temperature to consider negative feedback by TC-induced sea surface cooling on existing SST-based MPI. DMPI has significantly reduced the overestimation of maximum intensity of the existing SST-based MPI and has frequently been used to predict TC intensity and RI [2,19,32,33]. The variables based on intensification potential (POT; MPI − initial intensity) were the essential factors in the CSTIPS-DAT model. However, in this study, TC genesis was defined as the first moment of at least 35 kt intensity; thus, the

POT and DMPI had the same correlation coefficient. Because the current study focused on classifying the LMI of TCs at their genesis, the POT and DAT-based POT were excluded from the pool of potential variables. DATs and DMPIs had the highest correlation among all variables, reaching 0.54 and 0.56, respectively. OHC, a widely used index for upper-ocean thermal conditions, also had a high correlation coefficient (r = 0.52). Price [29] showed that OHC and DAT are well correlated in the high OHC range and deep water, but they are poorly correlated in low OHC and shallow continental shelves. Since the TWNP is mostly deep and has high OHC, the correlation coefficients of OHC and DAT are not very different there. All the variables were averaged from the genesis to 3.25 days along the TC track—the sum of the average time (1.7 days) and standard deviation (1.55 days) to reach LMI after TWNP TCs' occurrence.

Table 1. Potential variables in the present model and their correlation coefficients (r) with the lifetime maximum intensity for the 2004–2013 TWNP TCs (Cluster 2 in CTIPS-DAT). All the variables were averaged along the TC track from the genesis to 3.25 days.

Variable	Description	r
JDAY	The absolute value of Julian day—248	−0.27
LAT	Latitude of typhoon location	−0.33
LON	Longitude of typhoon location	0.07
SPD	Storm moving speed	−0.23
D200	Area-averaged (0 km to 1000 km) divergence at 200 hPa	0.05
RV500	Area-averaged (0 km to 1000 km) relative vorticity at 500 hPa	0.16
RV850	Area-averaged (0 km to 1000 km) relative vorticity at 850 hPa	0.04
U200	Area-averaged (200 km to 800 km) zonal wind at 200 hPa	−0.28
T200	Area-averaged (200 km to 800 km) air temperature at 200 hPa	−0.39
RHHI	Area-averaged (200 km to 800 km) relative humidity 500–300 hPa	0.32
RHLO	Area-averaged (200 km to 800 km) relative humidity 850–700 hPa	0.29
SH200	Area-averaged (200 km to 800 km) 200 hPa to 850 hPa vertical wind shear	−0.17
SH500	Area-averaged (200 km to 800 km) 500 hPa to 850 hPa vertical wind shear	−0.32
OHC	Area-averaged (0 km to 200 km) ocean heat contents	0.52
DAT10—DAT120	Ocean temperatures averaged from the near-surface down to the various depth (10 to 120 m, 10-m interval)	0.48–0.54
DMPI10—DMPI120	Maximum potential intensity using DAT10—DAT120	0.47–0.56

2.2.2. Classification and Regression Tree

The classification and regression tree (CART) is one of the decision-tree algorithms that are used for categorical and continuous variables [34]. The rules generated by the CART are easy to interpret, and overfitting can be avoided by postpruning a fully grown tree. The CART is a binary partitioning algorithm with only two child nodes from the parent node. The Gini index, the sum of the misclassification probabilities, can be used as an impurity or diversity measure in each node. It is expressed as follows:

$$G = 1 - \sum_{j=1}^{c} \left(\frac{n_j}{n}\right)^2 \tag{1}$$

where n is the number of observations in the node, c is the number of categories of target variables, and n_j is the number of observations belonging to the jth category of the target variable. The CART algorithm selects the best predictor to minimize the Gini index for each split and finds the optimal separation of each node, and this division process is repeated for each node to construct a decision tree. For example, in order to classify TC intensity using environmental variables, it is necessary to perform the classification by repeatedly changing the classification reference value (e.g., the sea surface temperature, 26 °C), to calculate the Gini index of the classified group, and to determine the optimal reference value which has a minimum Gini index. The above process is repeatedly performed as many times as the specified number of nodes. In this study, a classifier was developed based on the "fitctree" function included in Matlab's "statistics and machine learning toolbox".

2.2.3. The k-Fold Cross-Validation

The k-fold cross-validation is one of the most popular resampling techniques for increasing the statistical reliability of model performance measurements [35]. The procedure is as follows. First, the entire sample is divided into k equally sized subsamples in which one subsample is reserved as validation data. Second, the model is trained with $k - 1$ subsample, tested (or validated) with the retained subsample, and cross-validated k times until each subsample has been used for validation only once. Finally, the results of each step of the process are averaged to form an evaluation index, which can be used to perform forecast verification. The advantage of cross-validation is that all the cases are used for both training and validation, and each case is used for validation once. In this study, 10-fold cross-validation was used.

2.2.4. Synthetic Minority Oversampling Technique

When the binary classification model is trained with inequality data, a classifier will be biased toward the more frequently occurring class. The accuracy of the majority class is likely to be inflated in training, thus resulting in inappropriate predictive accuracy in testing. In the present study, the synthetic minority oversampling technique (SMOTE; [36]), one of the most commonly used oversampling techniques, was used to avoid the inequality sample problem. It randomly extracts samples from the minority class and increases the number of samples by generating synthetic samples with the ambient values of the extracted samples. In this study, the number of nearest neighbors to consider was set to five.

3. Results

The 2004–2016 distribution of LMI in the TWNP had two local maxima at approximately 50 kt and 100 kt, and a local minimum at 70 kt (Figure 2a). A bimodal distribution of the relative frequency of the LMI was also found in the WNP (Figure 2b). However, unlike the TWNP, the first peak in the WNP was higher than the second. The TWNP is a sub-basin in which the strongest TCs in the WNP occur, so the relative frequency of the strong TCs (2nd peak) was higher than that of the weak TCs (1st peak). In this study, the TWNP TCs were classified into two types: those with LMI above 70 kt (strongly developing TCs; A70) and those with LMI below 70 kt (weak TCs; B70).

Intensity prediction using the CSTIPS-DAT [2] revealed large mean absolute error (MAE) values and bias for the two classified groups, A70 and B70 (Figure 3a). As predicted, this was because the model was trained with the entire TWNP TCs that contain both weakly and strongly developing TCs, resulting in a negative bias (underestimation; see the red solid line in Figure 3a) for A70 and a positive bias (overestimation; see the blue solid line in Figure 3a) for B70. Indeed, most of the MAE values in the TWNP were related to the large biases, suggesting that the bias correction using individual models for A70 and B70 reduced the intensity prediction error. Overall, the MAE and bias were greater in B70 than in A70. This was related to the fact that during training, the model fit A70 better than B70 because A70 had about four times more samples than B70. In fact, the numbers of samples of the A70 and B70 groups were 60 and 17 TCs, respectively, during the training period, and 26 and 10 TCs, respectively, during the test periods (Table 2). To resolve the inequality in the training data set, SMOTE was used to increase the number of samples for B70 to 60, as in A70.

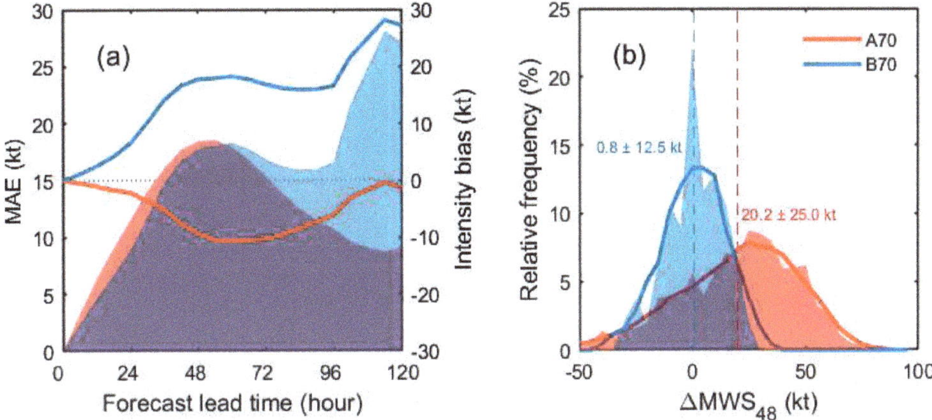

Figure 3. Comparison of (**a**) mean absolute errors (shading) and biases (solid lines) for intensity predictions of A70 (TCs with LMI greater than 70 kt) and B70 (TCs with LMI less than 70 kt) at each lead time for the 2013–2014 TWNP TCs. (**b**) Comparison of the relative frequencies of intensity change in 48 h in the classified groups (red: A70; blue: B70). The shaded areas show the raw data binned into 5 kt bins. The thick lines are the smoothed relative frequencies with a window width of 15 kt. The dashed lines indicate the means of each group. The mean values with ±σ are represented in colored text.

Table 2. Number of A70 and B70 tropical cyclones in 2004–2013, 2014–2016, and 2004–2016.

Period	A70	B70	Total
2004–2013	60	17	77
2014–2016	26	10	36
2004–2016	86	27	113

Figure 3b compares the relative frequencies of the intensity change for A70 and B70. The mean intensity change within 48 h was 20.2 ± 25.0 kt for A70 and 0.8 ± 12.5 kt for B70. The two-tailed Student's *t*-test revealed that the difference between the means of the two groups was statistically significant at the 5% test level. Therefore, it was expected that an LMI-based classification could reduce the variance of the intensity change in Cluster 2 and that the intensity prediction would be improved by the development of specific prediction models for each intensity type.

A confusion matrix [37] was used to calculate verification measures, namely the probability of detection (POD), false alarm rate (FAR), and accuracy. The POD is the ratio of the number of times a correct warning is issued for a target event to the total number of target events. The FAR is the number of times a warning is issued but an event does not occur divided by the number of times the warning is issued. The POD, FAR, and accuracy were calculated as follows:

$$POD = \frac{TP}{TP + FN} \tag{2}$$

$$FAR = \frac{FP}{FP + TP} \tag{3}$$

$$Accuracy = \frac{TP + TN}{TP + FP + TN + FN} \tag{4}$$

where *TP* is the true positive, *TN* is the true negative, *FP* is the false positive, and *FN* is the false negative. In this study, A70 was defined as the target class.

A decision tree generates the rule until the number of samples in a leaf drops below a specified size, i.e., the minimum leaf (min-leaf) size. The min-leaf size determines when splitting should be stopped; therefore, it is an important parameter that needs to be carefully tuned. Figure 4a presents the classification performance of the decision tree during the training period with various min-leaf sizes. The skill scores can be used to set the parameters. Naturally, the highest accuracy and POD were achieved at the min-leaf size of 1, and the performance score decreased with increased min-leaf sizes. The FAR varied by 0–12% with the min-leaf size; however, no significant trend was associated with the min-leaf size.

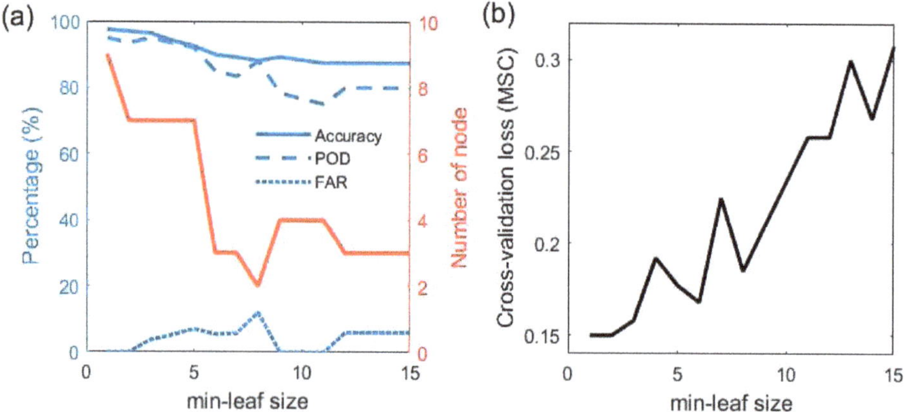

Figure 4. (**a**) Skill scores (blue lines) and the number of nodes (red line) at each minimum leaf size used in the decision-tree algorithm. (**b**) Distribution of cross-validation loss (mean misclassification rate) on the basis of the minimum leaf size using the *k*-fold cross-validation method.

A decision tree with a smaller min-leaf size usually has better performance. However, a small min-leaf size generates a complicated tree with many nodes, making a physical interpretation difficult. In addition, complicated trees can cause overfitting problems in classifications with insufficient sample sizes. A model should be trained to make reliable predictions for the test data. Overfitting is the result of modeling with noise instead of the underlying relationship. An excessively overfitted model performs poorly in real-time predictions because it is tuned to overreact to minor fluctuations in the training data. To avoid the prediction instability of overfitting, we determined the optimal min-leaf size using comparisons of the cross-validation (CV) loss. In this study, *k*-fold cross-validation was used to obtain the CV loss by averaging the misclassification rate (MSC), as shown below:

$$CV = \frac{1}{k} \sum_{i=1}^{k} MSC_i \tag{5}$$

$$MSC_i = \frac{n_{miss,\ i}}{n_i} \tag{6}$$

where *k* is the number of fold (here is set to be 10), $n_{miss,i}$ is the number of misclassification samples in *i*th test set, and n_i is the total number of samples in *i*th test set.

Figure 4b shows the change of the CV loss with min-leaf sizes. The CV loss tended to increase as the min-leaf size increased. The CV was the smallest at min-leaf sizes of 1 and 2, followed by local minima at 6 and 8. Min-leaf sizes of 1 and 2 required nine and seven nodes (fairly complex structure), respectively, and min-leaf sizes of 6 and 8 required three and two nodes, respectively (red line in Figure 4a). In this study, the min-leaf size was set to 6 to make the decision tree structure relatively simple with a small CV loss.

The trained decision tree included three nodes with four decision rules. Table 3 lists the decision rules governing the decision tree. Rule 1 shows that it is difficult for a TC in a low DMPI20 environment to intensify as it develops. MPI has been the most critical predictor in previous statistical intensity prediction models [38–42]. In Rule 1, shallow (i.e., 20 m deep) DMPI was selected as a classification factor, and this informed the classification of many weak TCs. Weak TCs cannot interact with the deep ocean; thus, the shallow-depth ocean-temperature-based MPI can be a good criterion for categorizing weak TCs.

Table 3. Description and the confidence of the rule of the developed decision tree. Note that all variables here were averaged along the TC track from genesis to 3.25 days.

Rule NO.	Decision Rules	The Confidence of the Rule
1	If DMPI20 < 114 kt, then TC will not develop above 70 kt.	45/51 = 88.2%
2	If DMPI20 ≥ 114 kt and LAT ≥ 22.1° N, then TC will not develop above 70 kt.	8/9 = 88.9%
3	If DMPI20 ≥ 114 kt, LAT < 22.1° N and DAT100 < 26.3 °C, TC will not develop above 70 kt.	4/6 = 66.7%
4	If DMPI20 ≥ 114 kt, LAT < 22.1° N and DAT100 ≥ 26.3 °C, TC will develop above 70 kt.	51/54 = 94.4%

Rule 2 states the following: If DMPI20 ≥ 114 kt and LAT ≥ 22.1° N, TCs are less likely to intensify to more than 70 kt. This suggests that it is difficult for a TC that stays at high latitudes on average during development to be classified as A70. This is because TCs with higher LAT tend to move northward and thus their tracks become closer to the polar westerlies, resulting in increased vertical wind shear that suppress TC intensification. The selection of LAT explains why vertical wind shear (SH200 and SH500), a well-known dynamic index related to TC intensity, was not singled out in the rule.

Rule 3 states the following: If DMPI20 ≥ 114 kt, LAT < 22.1° N, and DAT100 < 26.3 °C, a TC cannot intensify to more than 70 kt. This suggests that a high DMPI20 and a low LAT are favorable for intensity; however, TCs are less likely to develop with strong intensity if DAT100 is less than 26.3 °C. Price [21] suggested that 100 m is the typical vertical mixing depth that major TCs induce; thus, DAT100 is the realistic temperature that represents the sea surface thermal conditions under intense TCs. If DAT100 is less than 26.3 °C, which is close to the 2 m dew point temperature of the tropics [43], the ocean can no longer supply heat to the TC, thus reducing the likelihood of strong intensification.

Rule 4 states the following: If DMPI20 ≥ 114 kt, LAT < 22.1° N, and DAT100 ≥ 26.3 °C, a TC can intensify to more than 70 kt. This rule suggests that the development of intense TCs generally occurs when all three conditions are satisfied. The confidence of this rule was 94.4%.

To evaluate the capability of the decision tree to classify intensity, we analyzed the accuracy during the training and test period. The results showed a classification accuracy of 90.0% for training (Table 4) and 80.5% for testing (Table 5). According to the confusion matrix for the test period (Table 5), 24 of 26 TCs were correctly classified as A70, and 5 of 29 that were classified as A70 were B70. Thus, the POD had 92.3%, and the FAR had only 17.2%. These results exhibited high enough accuracy to build an independent statistical model for the TC groups classified on the basis of this algorithm.

Table 4. Confusion matrix for the training period.

		Model	
		A70	B70
Observed	A70	51	9
	B70	3	57

Table 5. Confusion matrix for the test period.

		Model	
		A70	B70
Observed	A70	24	2
	B70	5	5

4. Discussion

Kim et al. [2] classified TCs on the basis of their track patterns, by which the intensity characteristics could be classified. They showed that the prediction performance could be improved by reducing the variance of the predictand through the development of an individual model for each cluster. This study attempted to further reduce the predictand variance on the basis of the LMI classification, especially for Cluster 2 (TWNP TCs) of CSTIPS-DAT. The TWNP TCs show a bimodal LMI distribution, which can be classified as weakly (B70) and strongly developing TCs (A70). Because of this bimodality, the intensity prediction estimated using the CSTIPS-DAT showed large MAEs for the two groups. The large MAEs are mostly attributed to significant positive and negative biases for B70 and A70, respectively. This implies that correcting the biases through binary classification and developing independent prediction models for the classified groups can reduce the predictand variance and ultimately improve TC intensity prediction.

To improve the performance of the CSTIPS-DAT and to increase the understanding of LMI bimodality, this study developed a CART-algorithm-based decision tree which classifies the TC type at the time of genesis, based on whether or not it will reach an intensity of 70 kt or more during its lifetime. Among the 38 potential predictors, CART selected three variables that reached an accuracy of 90.0% in the training period (2004–2013) and 80.5% in the testing period (2014–2016). The selected variables were DMPI20, LAT, and DAT100. The splitting values were 114 kt for DMPI20, 22.1° N for LAT, and 26.3 °C for DAT100. The four developed rules indicate that the prestorm ocean thermal conditions (DMPI20 and DAT100) and latitude play a key role in determining the LMI in the TWNP.

It should be noted that DAT100 played an essential role in the decision tree developed for binary classification. For the unclassified TWNP TCs (black line in Figure 5), the correlation coefficients between various DATs and LMI were highest at DAT50. However, for strongly developing TCs (red line in Figure 5), the correlation was highest in DAT100. Price [21] proposed DAT100 as an oceanic index reflecting the sea surface cooling induced by Saffir–Simpson Category 3 TCs (96–113 kts). Interestingly, the Category 3 intensity belonged to the second peak of the LMI distribution (Figure 2a) and accounted for about 40% of the TWNP TCs. In contrast, for weak TCs (blue line in Figure 5) the correlation was very low at all DATs. This suggests that the pre-existing ocean thermal structures along the track are not essential in determining the LMI for weak TCs. Again, this highlights the need to develop individual models that consider key environmental factors differently depending on the classified groups.

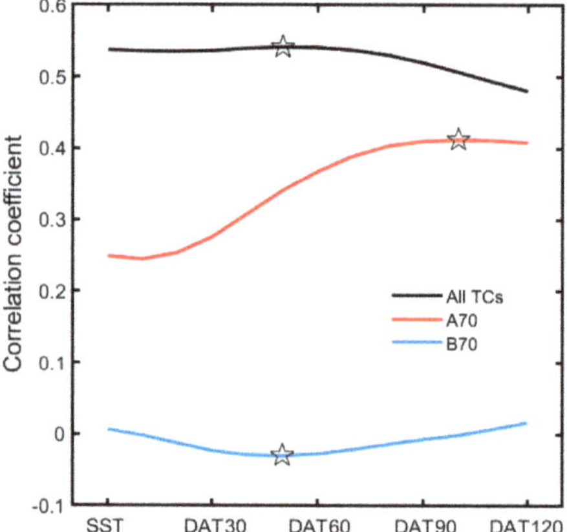

Figure 5. Comparison of correlation coefficients between various DATs and LMI for A70, B70, and all TCs. Black, red, and blue lines indicate unclassified TWNP TCs (i.e., all TCs), A70, and B70, respectively. Open stars represent the locations with the maximum values for each group.

5. Conclusions

Understanding the bimodal LMI distribution is important for improving TC intensity prediction. Previously known causes of this bimodality are the reduction of air–sea roughness at a particular wind speed range [10] and the presence or absence of rapid intensification events [8]. However, due to the lack of observational data in extreme winds, it is still difficult to fully understand the cause of the bimodal distribution. This study cannot directly explain the mechanism of the bimodality with the rules discovered, but it does present environmental parameters and their thresholds that can distinguish the two modes. This will make some contribution to a better understanding of the causes of the bimodal LMI distribution.

In this study, the CART algorithm, a machine-learning algorithm, was used for classification. Although the CART algorithm is widely used for binary classification, it cannot be affirmed that it is the optimal classification algorithm for classification of intensification types. Therefore, as in previous studies [22,25] that compared and evaluated several machine-learning algorithms for binary classification, research to find the optimal classification method by applying new classification tools must be conducted.

Author Contributions: Conceptualization, S.-H.K. and I.-J.M.; methodology, S.-H.K.; validation, S.-H.K., H.-W.K. and S.K.K.; formal analysis, S.-H.K.; data curation, S.-H.W.; writing—original draft preparation, S.-H.K.; writing—review and editing, S.-H.K., I.-J.M., S.-H.W., H.-W.K. and S.K.K.; supervision, I.-J.M., S.K.K. and S.-H.W. All authors have read and agreed to the published version of the manuscript.

Funding: This research was a part of the project titled 'Study on Air-Sea Interaction and Process of Rapidly Intensifying Typhoon in the Northwestern Pacific', funded by the Ministry of Oceans and Fisheries, Korea. This work was supported by the National Typhoon Center at the Korea Meteorological Administration ('Development of typhoon analysis and forecast technology', KMA2018-00722) and by Basic Science Research Program through the National Research Foundation of Korea (NRF) funded by the Ministry of Education (2021R1A2C1005287).

Institutional Review Board Statement: Not applicable.

Informed Consent Statement: Not applicable.

Data Availability Statement: Not applicable.

Conflicts of Interest: The authors declare no conflict of interest.

References

1. DeMaria, M.; Sampson, C.R.; Knaff, J.A.; Musgrave, K.D. Is tropical cyclone intensity guidance improving? *Bull. Am. Meteorol. Soc.* **2014**, *95*, 387–398. [CrossRef]
2. Kim, S.-H.; Moon, I.-J.; Chu, P.-S. Statistical-dynamic typhoon intensity predictions in the western North Pacific using Track Pattern Clustering and Ocean Coupling Predictors. *Weather Forecast.* **2018**, *33*, 347–365. [CrossRef]
3. Emanuel, K.A. A statistical analysis of tropical cyclone intensity. *Mon. Weather Rev.* **2000**, *128*, 1139–1152. [CrossRef]
4. Park, D.-S.R.; Ho, C.-H.; Kim, J.-H. Growing threat of intense tropical cyclones to East Asia over the period 1977–2010. *Environ. Res. Lett.* **2014**, *9*, 014008. [CrossRef]
5. Moon, I.-J.; Kim, S.-H.; Klotzbach, P.; Chan, J.C.L. Roles of interbasin frequency changes in the poleward shifts of the maximum intensity location of tropical cyclones. *Environ. Res. Lett.* **2015**, *10*, 104004. [CrossRef]
6. Manganello, J.V.; Hodges, K.I.; Kinter, J.L., III; Cash, B.A.; Marx, L.; Jung, T.; Achuthavarier, D.; Adams, J.D.; Altshuler, E.L.; Huang, B.; et al. Tropical cyclone climatology in a 10-km global atmospheric GCM: Toward weather-resolving climate modeling. *J. Clim.* **2012**, *25*, 3867–3893. [CrossRef]
7. Zhao, M.; Held, I.M.; Lin, S.-J.; Vecchi, G.A. Simulations of global hurricane climatology, interannual variability, and response to global warming using a 50-km resolution GCM. *J. Clim.* **2009**, *22*, 6653–6678. [CrossRef]
8. Lee, C.-Y.; Tippett, M.K.; Sobel, A.H.; Camargo, S.J. Rapid intensification and the bimodal distribution of tropical cyclone intensity. *Nat. Commun.* **2016**, *7*, 10625. [CrossRef] [PubMed]
9. Torn, R.D.; Snyder, C. Uncertainty of tropical cyclone best-track information. *Weather Forecast.* **2012**, *27*, 715–729. [CrossRef]
10. Soloviev, A.V.; Lukas, R.; Donelan, M.A.; Haus, B.K.; Ginis, I. The air-sea interface and surface stress under tropical cyclones. *Sci. Rep.* **2014**, *4*, 5306. [CrossRef]
11. Bengtsson, L.; Botzet, M.; Esch, M. Will greenhouse-induced warming over the next 50 years lead to higher frequency and greater intensity of hurricanes? *Tellus* **1996**, *48A*, 57–73. [CrossRef]

12. Emanuel, K.A. Increasing destructiveness of tropical cyclones over the past 30 years. *Nature* **2005**, *436*, 686–688. [CrossRef] [PubMed]
13. Webster, P.J.; Holland, G.; Curry, J.A.; Chang, H.-R. Changes in tropical cyclone number, duration, and intensity in a warming environment. *Science* **2005**, *309*, 1844–1846. [CrossRef] [PubMed]
14. Zhang, W.; Gao, S.; Chen, B.; Cao, K. The application of decision tree to intensity change classification of tropical cyclones in western North Pacific. *Geophys. Res. Lett.* **2013**, *40*, 1883–1887. [CrossRef]
15. Kim, H.-S.; Kim, H.S. Development of scheme for tropical cyclone genesis using machine learning. In Proceedings of the Autumn Meeting, Busan, Korea, 31 October–2 November 2016; KMS: Seoul, Korea, 2016; pp. 847–848.
16. Quinlan, J. *C4.5: Programs for Machine Learning*; Morgan Kaufmann: Burlington, MA, USA, 1993; 302p.
17. Li, W.; Yang, C.; Sun, D. Mining geophysical parameters through decision-tree analysis to determine correlation with tropical cy-clone development. *Comput. Geosci.* **2009**, *35*, 309–316. [CrossRef]
18. Zhang, W.; Fu, B.; Peng, M.S.; Li, T. Discriminating developing versus non-developing tropical disturbances in the western North Pacific through decision tree analysis. *Weather Forecast.* **2015**, *30*, 446–454. [CrossRef]
19. Gao, S.; Zhang, W.; Liu, J.; Lin, I.-I.; Chiu, L.S.; Cao, K. Improvements in typhoon intensity change classification by incorporating an ocean coupling potential intensity index into decision trees. *Weather Forecast.* **2016**, *31*, 95–106. [CrossRef]
20. Park, M.S.; Kim, M.; Lee, M.I.; Im, J.; Park, S. Detection of tropical cyclone genesis via quantitative satellite ocean surface wind pattern and intensity analyses using decision trees. *Remote Sens. Environ.* **2016**, *183*, 205–214. [CrossRef]
21. Lee, H.-M.; Won, S.-H.; Cha, E.-J.; Jung, J.-U. Development of technique for tropical cyclone formation using machine learning. In Proceedings of the Autumn Meeting, Gyeongju, Korea, 30 October–1 November 2019; KMS: Seoul, Korea, 2019; pp. 544–545.
22. Kim, M.; Park, M.-S.; Im, J.; Park, S.; Lee, M.-I. Machine learning approaches for detecting tropical cyclone formation using satellite data. *Remote Sens.* **2019**, *11*, 1195. [CrossRef]
23. Nam, C.C.; Park, D.-S.R.; Ho, C.-H.; Chen, D. Dependency of tropical cyclone risk on track in South Korea. *Nat. Hazards Earth Syst. Sci.* **2018**, *18*, 3225–3234. [CrossRef]
24. Yang, R.; Tang, J.; Kafatos, M. Improved associated conditions in rapid intensifications of tropical cyclones. *Geophys. Res. Lett.* **2007**, *34*, L20807. [CrossRef]
25. Yang, R. A Systematic Classification Investigation of Rapid Intensification of Atlantic Tropical Cyclones with the SHIPS Database. *Weather Forecast.* **2016**, *31*, 495–513. [CrossRef]
26. Su, H.; Wu, L.; Jiang, J.H.; Pai, R.; Liu, A.; Zhai, A.J.; Tavallali, P.; DeMaria, M. Applying satellite observations of tropical cyclone internal structures to rapid intensification forecast with machine learning. *Geophys. Res. Lett.* **2020**, *47*, e2020GL089102. [CrossRef]
27. Wei, Y.; Yang, R. An Advanced Artificial Intelligence System for Investigating Tropical Cyclone Rapid Intensification with the SHIPS Database. *Atmosphere* **2021**, *12*, 484. [CrossRef]
28. Kaplan, J.; De Maria, M. Large-Scale Characteristics of Rapidly Intensifying Tropical Cyclones in the North Atlantic Basin. *Weather Forecast.* **2003**, *18*, 1093–1108. [CrossRef]
29. Price, J.F. Metrics of hurricane-ocean interaction: Vertically-integrated or vertically-averaged ocean temperature? *Ocean Sci.* **2009**, *5*, 351–368. [CrossRef]
30. Park, J.H.; Yae, D.E.; Lee, K.J.; Lee, H.J.; Lee, S.W.; Noh, S.; Kim, S.J.; Shin, J.Y.; Nam, S.H. Rapid decay of slowly moving typhoon Soulik (2018) due to interactions with the strongly stratified Northern East China Sea. *Geophys. Res. Lett.* **2019**, *46*, 14595–14603. [CrossRef]
31. Lin, I.-I.; Black, P.; Price, J.F.; Yang, C.Y.; Chen, S.S.; Lien, C.C.; Harr, P.; Chi, N.H.; Wu, C.C.; D'Asaro, E.A. An ocean coupling potential intensity index for tropical cyclones. *Geophys. Res. Lett.* **2013**, *40*, 1878–1882. [CrossRef]
32. Balaguru, K.; Foltz, G.R.; Leung, L.R.; D'Asaro, E.; Emanuel, K.A.; Liu, H.; Zedler, S.E. Dynamic Potential Intensity: An improved representation of the ocean's impact on tropical cyclones. *Geophys. Res. Lett.* **2015**, *42*, 6739–6746. [CrossRef]
33. Lee, W.; Kim, S.H.; Chu, P.S.; Moon, I.J.; Soloviev, A.V. An index to better estimate tropical cyclone intensity change in the western North Pacific. *Geophys. Res. Lett.* **2019**, *46*, 8960–8968. [CrossRef]
34. Breiman, L.; Friedman, J.H.; Olshen, R.A.; Stone, C.J. *Classification and Regression Trees*; Wadsworth & Brooks/Cole Advanced Books & Software: Monterey, CA, USA, 1984; ISBN 978-0-412-04841-8.
35. McLachlan, G.J.; Do, K.-A.; Ambroise, C. *Analyzing Microarray Gene Expression Data*; Wiley: Hoboken, NJ, USA, 2004.
36. Chawla, N.V. C4.5 and imbalanced data sets: Investigating the effect of sampling method, probabilistic estimate, and decision tree structure. In Proceedings of the International Conference on Machine Learning, Washington, DC, USA, 21–24 August 2003; International Machine Learning Society: Princeton, NJ, USA, 2003; Volume 3, pp. 66–73.
37. Powers, D.M. Evaluation: From precision, recall and f-measure to roc, informedness, markedness and correlation. *J. Mach. Learn. Technol.* **2011**, *2*, 37–63.
38. Knaff, J.A.; Sampson, C.R.; DeMaria, M. An operational Statistical Typhoon Intensity Prediction Scheme for the western North Pacific. *Weather Forecast.* **2005**, *20*, 688–699. [CrossRef]
39. Kaplan, J.; DeMaria, M.; Knaff, J.A. A revised tropical cyclone rapid intensification index for the Atlantic and eastern North Pacific basins. *Weather Forecast.* **2010**, *25*, 220–241. [CrossRef]
40. Gao, S.; Chiu, L.S. Development of statistical typhoon intensity prediction: Application to satellite observed surface evaporation and rain rate (STIPER). *Weather Forecast.* **2012**, *27*, 240–250. [CrossRef]

41. Kaplan, J.; Rozoff, C.M.; DeMaria, M.; Sampson, C.R.; Kossin, J.P.; Velden, C.S.; Cione, J.J.; Dunion, J.P.; Knaff, J.A.; Zhang, J.A.; et al. Evaluating environmental impacts on tropical cyclone rapid intensification predictability utilizing statistical models. *Weather Forecast.* **2015**, *30*, 1374–1396. [CrossRef]
42. Knaff, J.A.; Sampson, C.R.; Musgrave, K.D. An Operational Rapid Intensification Prediction Aid for the Western North Pacific. *Weather Forecast.* **2018**, *33*, 799–811. [CrossRef]
43. Wada, A. Reexamination of tropical cyclone heat potential in the western north pacific. *J. Geophys. Res. Atmos.* **2016**, *121*, 6723–6744. [CrossRef]

 *atmosphere*

Article

Reanalysis Product-Based Nonstationary Frequency Analysis for Estimating Extreme Design Rainfall

Dong-IK Kim [1,*], Dawei Han [1] and Taesam Lee [2,*]

1   Office 2.17, Queens Bulding, Water and Environment Research Group, Department of Civil Engineering, University of Bristol, Bristol BS8 1TL, UK; d.han@bristol.ac.uk
2   Department of Civil Engineering, ERI, Gyeongsang National University, Jinju 501, Korea
*   Correspondence: dk15461@bristol.ac.uk (D.-I.K.); tae3lee@gnu.ac.kr (T.L.)

Abstract: Nonstationarity is one major issue in hydrological models, especially in design rainfall analysis. Design rainfalls are typically estimated by annual maximum rainfalls (AMRs) of observations below 50 years in many parts of the world, including South Korea. However, due to the lack of data, the time-dependent nature may not be sufficiently identified by this classic approach. Here, this study aims to explore design rainfall with nonstationary condition using century-long reanalysis products that help one to go back to the early 20th century. Despite its useful representation of the past climate, the reanalysis products via observational data assimilation schemes and models have never been tested in representing the nonstationary behavior in extreme rainfall events. We used daily precipitations of two century-long reanalysis datasets as the ERA-20c by the European Centre for Medium-Range Weather Forecasts (ECMWF) and the 20th century reanalysis (20CR) by the National Oceanic and Atmospheric Administration (NOAA). The AMRs from 1900 to 2010 were derived from the grids over South Korea. The systematic errors were downgraded through quantile delta mapping (QDM), as well as conventional stationary quantile mapping (SQM). The evaluation result of the bias-corrected AMRs indicated the significant reduction of the errors. Furthermore, the AMRs present obvious increasing trends from 1900 to 2010. With the bias-corrected values, we carried out nonstationary frequency analysis based on the time-varying location parameters of generalized extreme value (GEV) distribution. Design rainfalls with certain return periods were estimated based on the expected number of exceedance (ENE) interpretation. Although there is a significant range of uncertainty, the design quantiles by the median parameters showed the significant relative difference, from -30.8% to 42.8% for QDM, compared with the quantiles by the multi-decadal observations. Even though the AMRs from the reanalysis products are challenged by various errors such as quantile mapping (QM) and systematic errors, the results from the current study imply that the proposed scheme with employing the reanalysis product might be beneficial to predict the future evolution of extreme precipitation and to estimate the design rainfall accordingly.

Keywords: Bayesian approach; nonstationarity; reanalysis products; quantile delta mapping

Citation: Kim, D.-I.; Han, D.; Lee, T. Reanalysis Product-Based Nonstationary Frequency Analysis for Estimating Extreme Design Rainfall. *Atmosphere* **2021**, *12*, 191. https://doi.org/10.3390/atmos12020191

Academic Editor: Alexandre Ramos
Received: 14 January 2021
Accepted: 27 January 2021
Published: 31 January 2021

1. Introduction

Design rainfall plays an essential role in planning a water-related infrastructure and it has been commonly estimated from the precipitation intensity–duration–frequency (IDF) relationship based on the historical records with the stationary assumption [1,2]. However, recent research has indicated that many regions over the world have experienced the pattern change of climatic extremes, especially for heavy rainfall [3,4]. Recalling that a water-related project is designed under stationary condition in practice, the temporal change of extreme rainfalls, so-called 'nonstationarity', may significantly affect the safety of the infrastructure. In other words, an increasing trend in heavy rainfall can underestimate the estimated future risk when an obvious trend exists in a target region. Note that the hypothesis of a non-stationary model can be altered by the presence of outliers with assuming parameters as random variables, while preserving the hypothesis of a stationary process [5].

The other major issue in rainfall frequency analysis is the lack of the gauge data resulting in significant errors in hydrological modellings. In a block maxima (BM) approach typically used in practical application, one generally collects annual maximum rainfalls (AMRs) from historical records to derive IDF relationships. However, in many regions including South Korea, the long-term meteorological record for a given catchment is largely limited to create the reliable IDF relationships. For instance, South Korea, with an area of approximately 100,032 km^2, has over hundreds of weather stations, but the number of stations with daily rainfalls over 40 years was at most a few dozen by the Korean War. Consequently, the rainfall quantiles are derived from less than 40-year data and necessarily contain significant uncertainties, which are associated with the sampling errors [6–10].

Furthermore, the use of external source is also critical to respect the ergodicity [11,12], since the properties of a stochastic process might be estimated only by limited observations. When the process is nonstationary, the ergodicity cannot hold and possible trends should be derived from external sources, different from observed time series.

Subsequently, the future risk in a region like South Korea can be relabily estimated by considering the nonstationary and extending the data length is critical. However, existing studies have explored individual aspects, but there have been no studies to combine two factors, nonstationarity and lack of data, at the same time. Thus, this study aims to reliably extend the data series, and then to explore the future risk change (i.e., nonstationarity) with the extended time series, which may depend on time, in South Korea.

To address the lack of data, we introduce long-term reanalysis datasets as substitutes of the observed. With their own assimilation techniques, reanalysis products have been globally provided with daily or sub-daily resolution and they have been applied in the global-, continental- and country-scale climate change analyses [13–18]. However, most reanalysis products provide the climate variables after the 1950s, and a few datasets assimilated by two representative institutes, the National Oceanic and Atmospheric Administration (NOAA) and the European Centre for Medium-Range Weather Forecasts (ECMWF), can cover the whole 20th century; the 20th century reanalysis (20CR) by the NOAA, and ERA-20c and ERA-20 cm by the ECMWF [15,18,19]. For ERA-20c and ERA-20cm by the ECMWF, they are based on the same simulation model, but ERA-20 cm does not consider observations in the assimilation process, unlike ERA-20c. For this reason, ERA-20c has a limitation in representing the actual synoptic situation [15–18]. Thus, we use two reanalysis datasets, 20CR and ERA-20c, produced by two different institutes as substitutes of the observed.

Like other model data, long-term reanalysis datasets include the systematic errors which vary with space [17,20]. Thus, the bias correction of reanalysis products, especially for regional-scale studies, should be performed before hydrologic applications. Numerous studies have suggested various bias correction methods from a delta change to a quantile mapping (QM) or multivariate scheme using a copula-based principle [17,21–24]. Each method has its pros and cons, but a QM method has been commonly used for precipitation due to its good performance [21,25–27]. We also improve the biases in the modelled by using a QM approach. In the BM principle, estimating IDFs are based on the AMRs of the data. Thus, the purpose of bias correction in this study is to collect the improved AMRs, and there are two main approaches for it; (1) to improve the whole rainfall distribution and derive the AMRs and (2) to directly correct the AMRs of the modelled. Compared with the former approach, the direct correction approach does not need to consider 'drizzling effect', which can impair the bias corrected values, and the errors in IDFs were less than that by improving whole wet-day data [2]. In this context, we collect the AMRs directly from the reanalysis products and then improve them with the QM approach.

To perform a nonstationary frequency analysis, it is essential to detect the significant long-term trends of the data. Previous studies for South Korea have documented that rainfalls have increased over time, especially during summer [28–34]. However, if we only focus on the 'AMRs', it is also true that there is no clear evidence for the trend in the AMRs of the observed [35]. Nadarajah and Choi [35] explored the trends of the AMRs for the observed in five stations over South Korea from 1961 to 2001, but there were no significant trends. For this reason, we assess the long-term trend of the AMRs for the bias-corrected from 1900 to 2010 as well as for the observed during the reference period (1974 to 2010). After detecting the significant trend, we estimate the design quantiles with the nonstationary return period.

To consider nonstationarity for rainfall or flood frequency analysis, numerous studies have commonly adopted a time-varying parameter scheme [36–41]. Conceptually, this approach assumes that a climate variable like daily rainfall has the same distribution function type, typically generalized extreme value (GEV) distribution, but the parameters are dependent on time. In this concept, a return period ($T_t = 1/(1 - F_z(z_{q0}, \theta_t))$) with a certain design quantile (z_{q0}) at time t can be easily derived from a cumulative distribution (F_z) with the time-varying parameters (θ_t) [42]. However, as estimating a quantile with a target period, for example 100-year, varies by time, the use of a "return period" concept can be meaningless under nonstationary condition [37,42,43].

However, as the notion of return period can still provide intuitive information to engineers, numerous studies have handled this issue and two different approaches have been proposed; the expected waiting time (EWT) approach and the expected number of exceedance (ENE) approach [42,44–47]. The former scheme, EWT, focuses on the 'expected waiting time' for the first occurrence exceeding the design rainfall (z_{q0}). On the other hand, the ENE approach obtains the target value by setting the expected number of exceedances over the design life T. Both concepts should be numerically solved to estimate the design quantile with a target return period under nonstationary condition. However, conceptually, the EWT method has a drawback that this concept needs infinite (or as long as possible) future exceedance probabilities, which can cause uncertainty [42,47]. For this reason, the ENE interpretation is considered for the nonstationary analysis in this study.

As aforementioned, the primary goal of this study is to explore design rainfall with non-stationary condition using the bias-corrected reanalysis products covering from 1900 to 2010. For this purpose, this study performs a three-step approach. First, we reduce the biases in ERA-20c and 20CR, especially for the AMRs, by using a QM approach. Secondly, we evaluate long-term trends of the AMRs for the observed and the reanalysis products to detect the nonstationarity. The final step in this study is to evaluate the design rainfalls under nonstationary condition and to compare them with the results by the classic observation-based estimation. This three-step analysis is based on the mainland of South Korea, and the detailed information on the study site and data used in this study is described in Section 2. The theoretical background for the methodology is introduced in Section 3. The results and discussion are summarized in Sections 4 and 5, respectively. Finally, the summary and conclusions are provided in Section 6.

2. Study Site and Data

2.1. Study Site and Local Gauge Data

This study is based on the mainland of South Korea, which lies between latitudes 34°–38.5° N and longitudes 126°–129.5° E, excluding all the islands including Jeju. The local gauge records are typically used for hydrologic applications, including frequency analysis. For South Korea, only dozens of weather stations can provide historical records for more than 40 years among hundreds of gauging stations. Thus, we obtained 48 in situ daily precipitation data, covering from 1974 to 2017, in the mainland of South Korea. After collecting the daily data from the Korea Meteorological Administration (http://www.kma.go.kr), we derived the annual maximum rainfalls (AMRs) for bias correction in all 48 stations. Note that, for St. 14 Andong, we ignored the AMRs from 1978 to 1982 due to the absence of records during the period. The daily AMR sequences were collected and compiled from the Korea Meteorological Administration (KMA). The coverage of the study area and the weather stations chosen in this study is shown in Figure 1, and the information for the stations is described in Table 1.

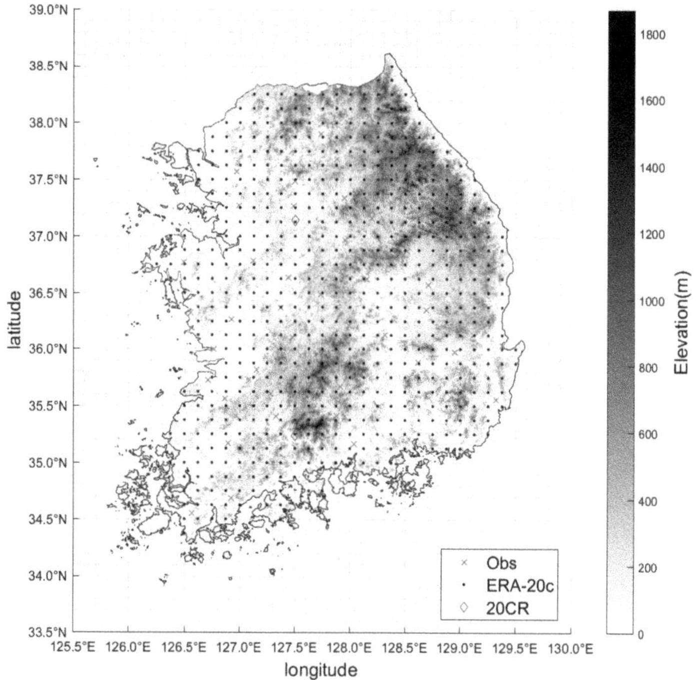

Figure 1. A map showing the study area, local gauging stations, grid points of ERA-20c and 20CR. The grey shading on the map indicates elevations.

Table 1. Station information employed in the study.

Station No.	Name	Latitude (°N)	Longitude (°E)	Elevation (m. asl)
St. 1	Sokcho	38.2508	128.5644	19.5
St. 2	Daegwallyeong	37.6769	128.7181	774.0
St. 3	Chuncheon	37.9025	127.7356	79.1
St. 4	Gangneung	37.7514	128.8908	27.4
St. 5	Seoul	37.5714	126.9656	11.1
St. 6	Incheon	37.4775	126.6247	69.6
St. 7	Wonju	37.3375	127.9464	150.0
St. 8	Suwon	37.2700	126.9875	38.3
St. 9	Chungju	36.9700	127.9525	116.5
St. 10	Seosan	36.7736	126.4958	30.3
St. 11	Cheongju	36.6361	127.4428	58.6
St. 12	Daejeon	36.3689	127.3742	70.3
St. 13	Chupungyeong	36.2197	127.9944	246.1

Table 1. *Cont.*

Station No.	Name	Latitude (°N)	Longitude (°E)	Elevation (m. asl)
St. 14	Andong	36.5728	128.7072	141.5
St. 15	Pohang	36.0325	129.3794	3.7
St. 16	Gunsan	36.0019	126.7631	24.6
St. 17	Daegu	35.8850	128.6189	65.5
St. 18	Jeonju	35.8214	127.1547	54.8
St. 19	Ulsan	35.5600	129.3200	36.0
St. 20	Gwangju	35.1728	126.8914	73.8
St. 21	Busan	35.1044	129.0319	71.0
St. 22	Mokpo	34.8167	126.3811	39.4
St. 23	Yeosu	34.7392	127.7406	66.0
St. 24	Jinju	35.1636	128.0400	31.6
St. 25	Yangpyeong	37.4886	127.4944	49.4
St. 26	Icheon	37.2639	127.4842	79.4
St. 27	Inje	38.0600	128.1669	201.6
St. 28	Hongcheon	37.6833	127.8803	142.3
St. 29	Jecheon	37.1592	128.1942	265.0
St. 30	Boeun	36.4875	127.7339	176.4
St. 31	Cheonan	36.7794	127.1211	24.0
St. 32	Boryeong	36.3269	126.5572	16.9
St. 33	Buyeo	36.2722	126.9206	12.7
St. 34	Geumsan	36.1056	127.4817	171.7
St. 35	Buan	35.7294	126.7164	13.4
St. 36	Imsil	35.6122	127.2853	249.3
St. 37	Jeongeup	35.5631	126.8658	46.0
St. 38	Namwon	35.4053	127.3328	91.7
St. 39	Jangheung	34.6886	126.9194	46.4
St. 40	Haenam	34.5533	126.5689	14.4
St. 41	Goheung	34.6181	127.2756	54.5
St. 42	Yeongju	36.8717	128.5167	212.2
St. 43	Mungyeong	36.6272	128.1486	172.0
St. 44	Uiseong	36.3558	128.6883	83.2
St. 45	Gumi	36.1306	128.3206	50.3
St. 46	Yeongcheon	35.9772	128.9514	95.0
St. 47	Geochang	35.6711	127.9108	222.4
St. 48	Sancheong	35.4128	127.8789	0.8

2.2. Reanalysis Products

In this study, we apply two representative century-long reanalysis products as ERA-20c by the ECMWF and 20CR by the NOAA. The ERA-20c was produced by assimilation technique using observations of surface pressure and surface marine winds only, and the product can globally cover the period from 1900 to 2010 with the spatio-temporally various resolutions [18]. On the other hand, 20CR, the first century-long reanalysis products by the NOAA, was assimilated with the ensemble Kalman filter technique using only surface pressure observations [19] and its latest version 2c could span the period from 1851 to 2014 with a spatial resolution of 1.875° × 1.9°. As the long-term climate records play an important role in the nonstationary analysis, we collect daily rainfall data from these two century-long reanalysis products in this study. More specifically, we extracted the daily precipitation records from 1900 to 2010, with the finest spatial resolution, 0.125° × 0.125° for ERA-20c and 1.875° × 1.9° for 20CR, respectively. From these daily rainfall series, this study derived the AMRs at grids in the mainland of South Korea from 1900 to 2010. The grid points over the sea were ignored in this analysis. The specific grid-scale points for ERA-20c and 20CR are shown in Figure 1. Note that only two grid points of 20CR covered the entire study region due to its low spatial resolution.

These reanalysis products are employed in a number of climatological studies [48–51]. Diro et al. [49] evaluated the spatial pattern of rainfall estimates for the reanalysis product of ERA-40 and noted that the product captures well the annual cycle over most of the country. Berntell et al. [48] investigated the strong multidecadal variability of the summer rainfall during the 20th century with reanalysis products and reported that the reanalysis product ERA-20CM shows the best representation of the multidecadal rainfall variability. Hua, et al. [51] assessed reanalysis products with quality-controlled radisonde observations and observed datasets to understand the rainfall climatology and variability over Central Equatorial Africa.

3. Methodology

3.1. Bias Correction

Although the century-long reanalysis precipitation data adopt the observed data when modeling, the modeled data still include the substantial biases. The bias correction approach should preliminarily be applied to the model values before further hydrologic applications. In the current study, we first carried out the bias correction by a quantile mapping (QM) approach, typically adopted in the bias correction studies [31,33,52,53]. Conceptually, the QM method reduces the errors by fitting a cumulative distribution of the modelled into that of the observed via a transfer function [21,53–55] (see Supplementary Material for detail).

For estimating the design rainfalls, BM approach using the AMRs is commonly adopted. To implement a reanalysis products-based BM approach, the bias-corrected AMRs should be collected and there are two approaches to collect them. Firstly, we can correct all wet-day precipitation data by QM methods and then, take the AMRs from the bias-corrected daily values. The other option is to directly improve the uncorrected AMRs by QM methods without considering the other rainfall data. If they are interested not only in the AMRs but also in all daily rainfalls, it would be better to apply the first option. However, as this study only focuses on exploring the non-stationary design rainfalls based on the BM, we utilize the latter option, which can reduce the error more efficiently than correcting the entire rainfall series [2]. To find out the most fitted transfer function, we applied three representative distributions, gamma, Gumbel, and GEV, for the bias correction of the AMRs, which are commonly employed in hydrologic application and extreme study [22,53,56,57]. As a QM approach is based on a one-to-one relationship, we matched 48 weather stations with the closest grid points of the ERA-20c and 20CR, and the bias-corrected values were collected at each station. Here, we assumed that the difference of spatial resolution between datasets can be ignored. One major issue in bias correction for a climate model data is how to correct the model values beyond the time range of the observation. The conventional approach of the QM algorithm was implemented with the assumption that the climate records are stationary for the whole projected period [18,28]. More specifically, the CDFs of the observed and the modeled are estimated using records for a reference period (i.e., historical period or calibration period), and the model data for the whole projected period are applied to this correction factor.

In this concept, the years from 1974 to 2010 were set as the reference period, while the whole period from 1900 to 2010 was considered as the projected period. For the stationary quantile mapping (SQM) scheme, there exist some extreme values beyond the range of the reference period, which may overestimate the bias-correction results, so an appropriate extrapolation scheme should be considered for them [2,25,33]. In the current study, we applied a constant extrapolation, which uses the correction values at the lowest and highest quantiles of the calibration range, suggested by Themeßl, Gobiet and Heinrich [25] to the events beyond the range of reference data, while the AMRs within the range were corrected by parametric approaches based on three different distributions, GEV, gamma, and Gumbel. Note that the SQM approaches with the GEV, gamma, and Gumbel were abbreviated as gevSQM and gamSQM and gumSQM, respectively.

One major problem in SQM approach is to ignore time-dependent characteristic such as a long-term trend. To handle this issue, several approaches have been tested, such as detrended quantile mapping and quantile delta approach [58,59]. Among these approaches, we applied the quantile delta mapping (QDM) method suggested by Cannon, Sobie and Murdock [31], because this approach can preserve the changes, not only in mean but also in extremes for the modelled data. In QDM, long-term trends in data are preserved by superimposing the relative change of quantiles between the reference period and the projected period, which are set to the same length. Thus, we first set the reference period to 1974–2010 as in SQM, and then divided the past projected period (1900–1973) into two periods, 1900–1936 and 1937–1973, to make the intervals equal to the reference period. Consequently, reanalysis daily precipitations were divided into three time periods with the same length (1900–1936, 1937–1973, 1974–2010) and the raw models in each time period were improved by the QDM principle [31,33] (see Supplementary Material for detail).

In this analysis, we compared the bias-corrected AMRs by QM approaches with the observations in 48 stations for the reference period (1974–2010). As the results by QDM approaches are identical to those by SQM schemes for the reference period, the QDM results were used to evaluate the performances of three different curves for the bias correction scheme.

3.2. Detecting Nonstationarity: Long-Term Trend Test

As aforementioned in the introduction section, the conventional approach for bias correction is based on the stationary condition for climate model records, but the real climate may follow the non-stationary feature in terms of century-long trend. To find out the significance of the AMR trend over South Korea, we evaluated the long-term trends of the observation and the bias corrected reanalysis data. For the trend test, a non-parametric method, the Mann–Kendall test was applied in this study. The significance of trends was evaluated by comparing the test statistic Z with the standard normal variate at the desired significance [60]. When $|Z| > Z_{1-\alpha/2}$ for the standard normal deviate $Z_{1-\alpha/2}$ with the significance level α (=0.05 in the current study), the null hypothesis is rejected and a significant trend in a time series. For the slope, the Theil–Sen approach [61–64] defined by the median among the ranked slope estimates is applied (see Supplementary Material for detail).

We first analyzed the trends of the AMRs taken from both the observation and the bias-corrected reanalysis data for the reference period (1974–2010). To estimate nonstationarity over the 20th century, the century-long trends of the bias-corrected AMRs from 1900 to 2010 were also detected.

3.3. Rainfall Frequency Analysis with Nonstationary Condition

In hydrological models, time-varying parameter schemes have been commonly adopted for non-stationarity analysis in hydrometeorological applications [38,40–42,44,65]. As GEV family is typically applied for estimating IDFs in practice, we applied a GEV distribution with the time-varying location parameter (μ_t), while scale (σ) and shape (ξ) parameters were set as constant. The location parameter is assumed as a time-depending linear function, and under the nonstationary condition.

To quantify the parameters for the GEV curve under nonstationary condition, we apply the Bayesian principle suggested by Cheng and AghaKouchak [1]. In this scheme, numerous parameter sets are estimated from the joint posterior distribution using the differential evolution Markov chain (DE-MC), which is based on the genetic algorithm differential evolution for global optimization with the Markov chain Monte Carlo (MCMC) principle [1].

With the time-varying parameter chains, the next step is to estimate the return period for a given design quantile under nonstationary condition. Numerous studies have dealt with the non-stationary interpretation, and there were two main approaches to handle it as (1) the expected waiting time (EWT) method and (2) the expected number of exceedance (ENE) method [42,44,45,47,66].

Both EWT and ENE are applicable for nonstationary events. However, as stated in the Introduction section, the EWT approach has a drawback of requiring infinite (or as long as possible) future exceedance probabilities in order to numerically solve the problem [42,47]. For this reason, we applied the ENE interpretation for estimating the design quantile with the return period from 10-year to 200-year (see Supplementary Material for more detail) (Figure 2).

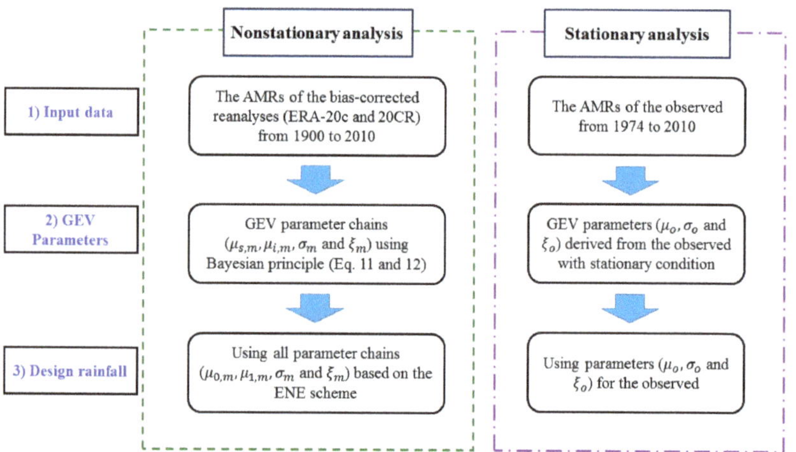

Figure 2. A flow chart for estimating design rainfall with the nonstationary condition and stationary condition (see Supplementary Material for the detailed equations).

4. Results

4.1. Bias Correction

To improve the uncorrected AMRs of ERA-20c and 20CR, we applied QM approaches based on three different distributions, gamma, Gumbel and GEV. The overall bias-corrected values over 48 stations for the reference period (1974–2010) were assessed by RMSE (mm) and NSE, as illustrated in Figure 3 and Table 2. Conceptually, the bias-corrected values by QDM and SQM methods with the same distribution are identical for the reference period. Thus, the outputs by three different distributions were denoted as gevQM, gamQM and gumQM. The overall comparison between the observed and the modelled indicated the significant reduction of errors in all QM schemes. The result showed that GEV distribution performed the best in both ERA-20c and 20CR. The bias-corrected ERA-20c by gevQM had 17.56 mm for RMSE and 0.924 for NSE, while the values by gamQM and gumQM were from 22.34 mm to 26.08 mm for RMSE and from 0.831 to 0.876 for NSE. For 20CR, the model efficiency by gevQM with 20.63 mm for RMSE and 0.894 for NSE dominated those by gamQM and gumQM with from 22.46 mm to 26.86 mm for RMSE and from 0.821 to 0.875 for NSE.

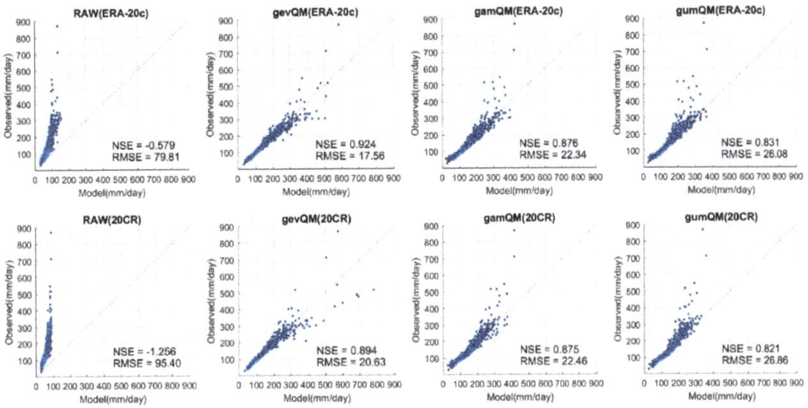

Figure 3. Scatter plots between the annual maximum rainfalls (AMRs) of the observation and the model data (the raw reanalyses (RAW(ERA-20c) and RAW (20CR)) and the bias-corrected reanalyses (i.e., ERA-20c and 20CR) by the quantile mapping (QM) approaches with generalized extreme value (GEV), gamma and Gumbel distributions (gevQM, gamQM and gumQM)) over 48 stations, from 1974 to 2010.

Table 2. Error estimation results of RMSE (mm) and NSE for the uncorrected (RAW) ERA-20c and 20CR, and the bias-corrected reanalyses (i.e., ERA-20c and 20CR) by the QM approaches with GEV, gamma and Gumbel distributions (gevQM, gamQM and gumQM) over 48 stations from 1974 to 2010.

Method	ERA-20c		20CR	
	RMSE (mm)	NSE	RMSE (mm)	NSE
RAW	79.81	−0.579	95.40	−1.256
gevQM	17.56	0.924	20.63	0.894
gamQM	22.34	0.876	22.46	0.875
gumQM	26.08	0.831	26.86	0.821

To spatially evaluate the performance, we also implemented the error estimation in individual stations, as shown in Figures 4 and 5 for NSE and RMSE, respectively. Figure 4 illustrates NSE values for the AMRs of the bias-corrected ERA-20c and 20CR based on QM approaches (i.e., gevQM, gamQM and gumQM) in 48 stations, whereas Figure 5 indicates RMSE values. The model efficiencies were generally over 0.8 for NSE, in all model values except a few stations. For RMSE, the most error estimates were less than 30 mm, which indicates the significant reduction of the bias. To clearly analyze the range of the model values, we used a boxplot scheme based on the individual error estimates in 48 stations, as illustrated in Figure 6, and compared the mean values of the error estimation results in individual stations as described in Table 3. The boxplot for NSE in Figure 6a indicated that the median values were over 0.9 in all QM approaches and most values were within the range from 0.6 to 1. Especially, gevQM for ERA-20c showed the best efficiencies among three QM schemes, whereas for 20CR, gevQM and gamQM had a better performance than gumQM. The analysis on RMSE also showed the similar result (Figure 6b). The median values were generally within 10 to 15 mm in all bias-corrected values and gevQM for ERA-20c performed the best and gevQM for 20CR and gamQM for ERA-20c and 20CR closely followed. In terms of the mean, gevQM for ERA-20c showed the best efficiencies, with 14.30 mm for RMSE and 0.933 for NSE as described in Table 3. For 20CR, NSE values for gevQM and gamQM were similar but RMSE for gevQM, 16.69 mm, was slightly smaller than that of gamQM, 17.48 mm. These results suggest that the applied QM approaches can significantly reduce the error in the AMRs of reanalyses (i.e., ERA-20c and 20CR), and among three different transfer functions, GEV distribution could be the best option for bias correction of the AMRs, especially for ERA-20c.

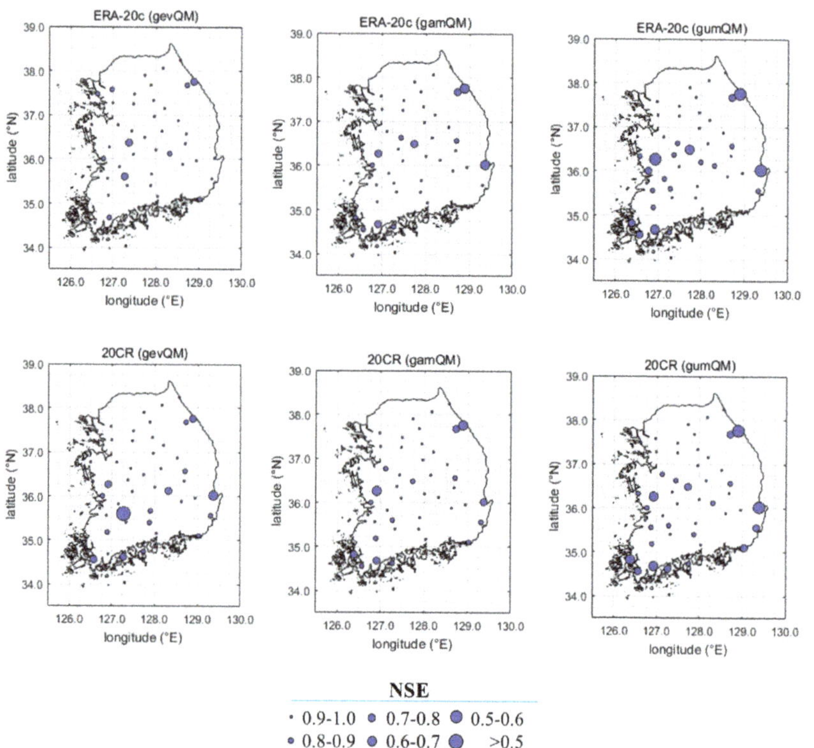

Figure 4. Mapping Nash–Sutcliffe efficiency (NSE) for the AMRs of the bias corrected ERA-20c (above) and 20CR (bottom) by QM approach (gevQM) in 48 stations for the reference period (1974–2010).

Table 3. Mean of error estimation results (RMSE (mm) and NSE) for the AMRs of the bias-corrected ERA-20c and 20CR by the QM approaches with GEV, gamma and Gumbel distributions (gevQM, gamQM and gumQM) in 48 stations from 1974 to 2010.

Method	ERA-20c		20CR	
	RMSE (mm)	NSE	RMSE (mm)	NSE
gevQM	14.30	0.933	16.69	0.905
gamQM	17.29	0.909	17.48	0.907
gumQM	20.31	0.871	21.09	0.864

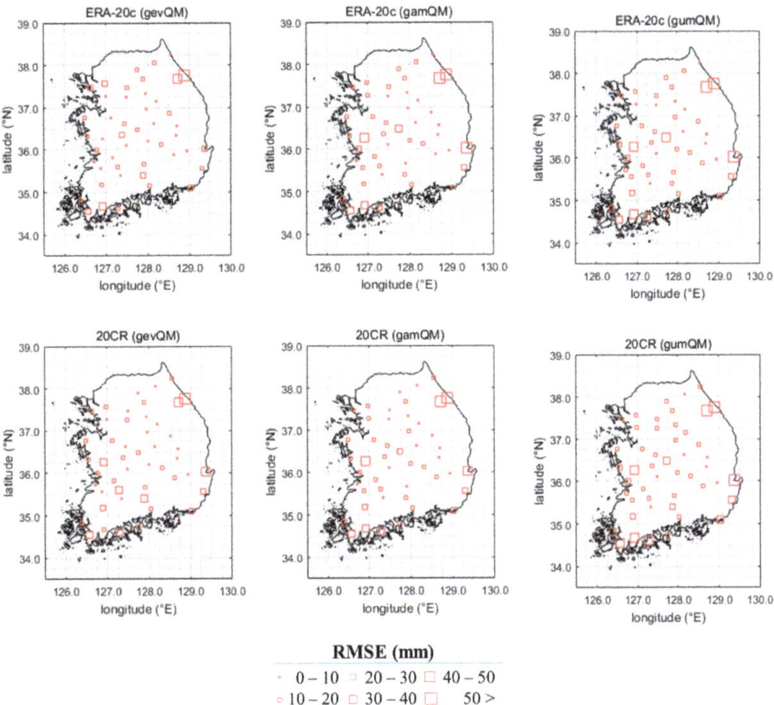

Figure 5. Root mean square error (RMSE) for the AMRs of the bias corrected ERA-20c (above) and 20CR (bottom) by QM approach (gevQM) in 48 stations for the reference period (1974–2010).

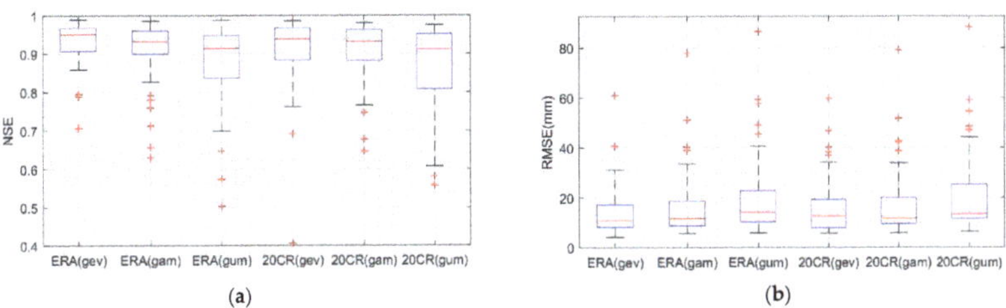

Figure 6. Boxplots of (**a**) NSE and (**b**) RMSE (mm) results for the AMRs of the bias corrected ERA-20c (ERA) and 20CR by QM approaches using three different distributions, GEV (gev), gamma (gam), and Gumbel (gum) in 48 stations.

4.2. Long-Term Trend

To consider the nonstationary condition in rainfall frequency analysis, the long-term trend of the AMRs should be necessarily detected. For this purpose, we analyzed the long-term trend of AMRs for the observed (Obs) and the bias corrected for the reference period (i.e., 1974–2010) using a non-parametric method, Mann–Kendall test as shown in Figure 7. To further evaluate the observed trend, we additionally analyzed the AMRs of the observation for the extended period, from 1974 to 2017 (Obs0). The results in both Obs and Obs0 had no significant trend at the 0.05 significant level except a few stations. More specifically, among 48 stations, only 5 stations for Obs and 2 stations for Obs0 presented significant trend, respectively. The performances for the bias-corrected reanalyses showed similar results. The AMRs of the bias-corrected reanalysis data (ERA-20c and 20CR) by QM approaches as well as the raw values (RAW) had no significant trends at the 0.05

significance level in all comparisons. Note that the recent 40 years' data do not illustrate any significant data in the observation, as well as in the reanalysis data.

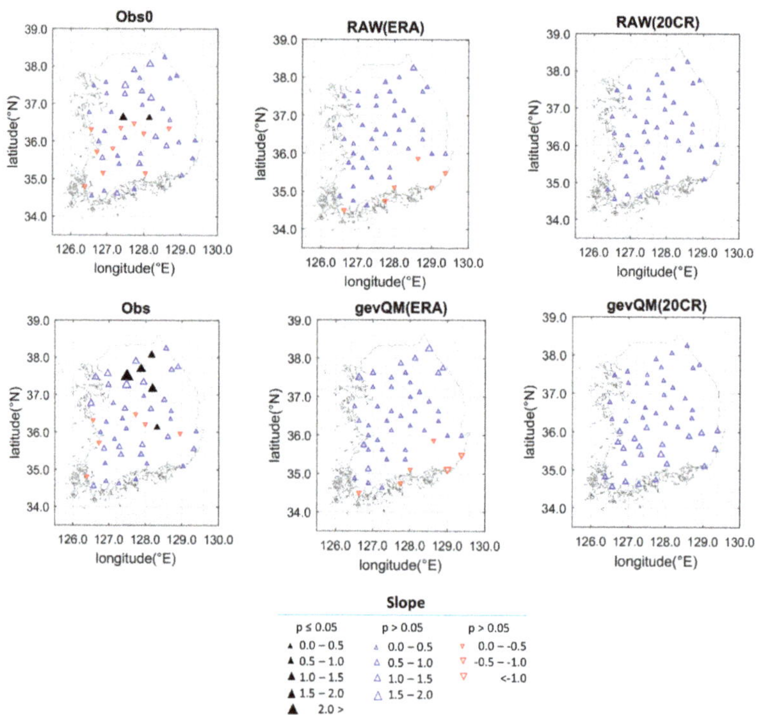

Figure 7. Trends in the AMRs of the observation for 1974–2017(Obs0) and 1974–2010(Obs), respectively, and the AMRs of the raw reanalyses (RAW) and the bias-corrected values by the QM approach (gevQM) for the reference period (1974–2010). ERA and 20CR mean the data from the ERA-20c and 20CR in this figure. Note that solid triangles represent significant trends at the 95% confidence levels, while hollow triangles mean no significant trends. The upward-pointing triangle and downward-pointing indicate increasing slope and decreasing slope, respectively, whereas the size of triangle represents the magnitude of trends.

To estimate nonstationarity over the 20th century, we also checked a century-long trend using the bias corrected AMRs of ERA-20c and 20CR from 1900 to 2010, instead of the observation periods Obs and Obs0. Figure 8 illustrates the trend test results for the AMRs of the bias corrected reanalyses in 48 stations from 1900 to 2010. For ERA-20c, the corrected values by SQM approaches (gevSQM, gamSQM and gumSQM) had obvious increasing trends in all stations, and the QDM algorithms (gevQDM, gamQDM and gumQDM) also indicated the significant increasing trends at 0.05 significance level, except a few points. The results for 20CR were similar to those for ERA-20c. The bias-corrected values demonstrated the obvious increasing trends for both SQM and QDM schemes of the whole 20th century. These results imply that the AMRs in South Korea may have an increasing trend over the 20th century, although the AMRs in recent decades are not able to clarify the nonstationary characteristic.

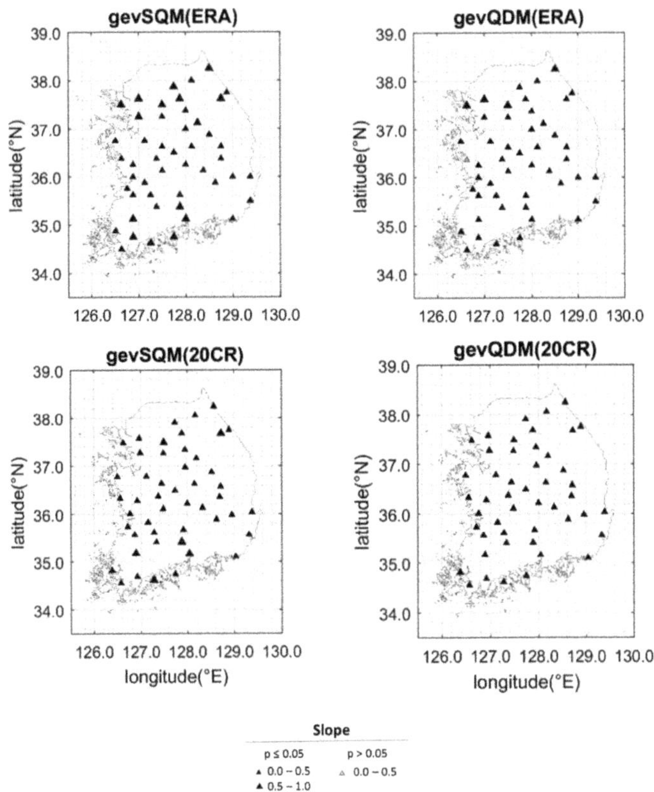

Figure 8. The trends in the AMRs of the bias corrected ERA-20c and 20CR by QM approaches as in Figure 7, but the data period is covered from 1900 to 2010.

To evaluate the magnitude of the trends, inter-annual variabilities over 48 stations from 1900 to 2010 were also analyzed as illustrated in Figure 9. Individual values for each station are presented with thin weak blue and red lines for ERA-20c and 20CR, respectively, while the means of 48 stations are presented with thick strong blue and red lines. Although there was the difference in specific movements between ERA-20c and 20CR, the trends of the overall means for 48 stations represented by bold blue color for ERA-20c and bold red color for 20CR illustrate the significant increasing trends for both cases.

The slopes for ERA-20c were within the range from 0.40 mm/year to 0.55 mm/year, whereas 20CR had fewer slopes from 0.34 to 0.45 mm/year as described in Table 4. In comparison between SQM and QDM approaches, the trends by QDM schemes were lower than those by the corresponding SQM methods in both ERA-20c and 20CR. With the assumption that two reanalyses, ERA-20c and 20CR, could substantially reproduce the long-term trend, these temporal patterns imply the clear non-stationarity of the AMRs in South Korea from 1900 to 2010.

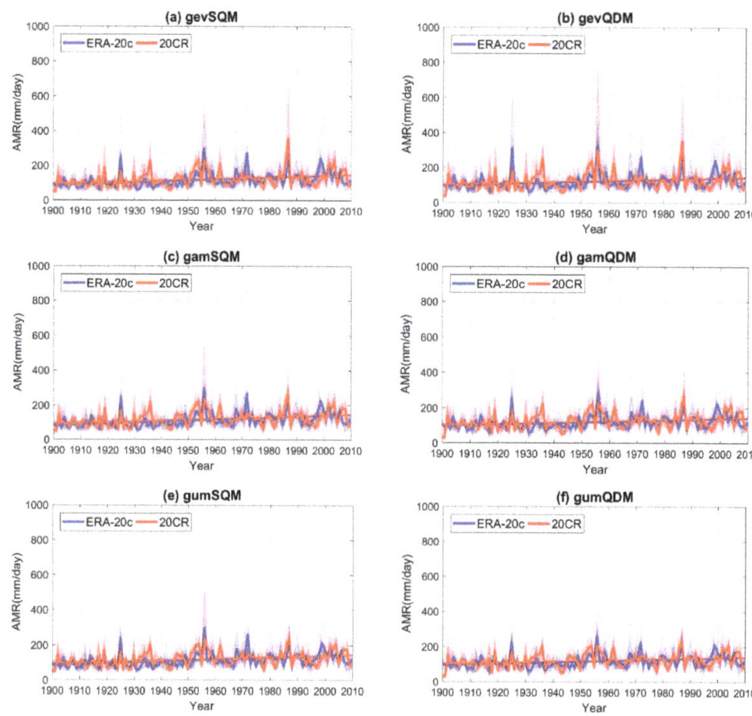

Figure 9. Inter-annual change of the AMRs of the bias corrected ERA-20c and 20CR by QM approaches ((**a**) gevSQM, (**b**) gevQDM, (**c**) gamSQM, (**d**) gamQDM, (**e**) gumSQM, and (**f**) gumQDM]. The bold blue line and bold red line represent the means of the ERA-20c AMRs and the 20CR AMRs over 48 stations, respectively, while the light blue and red lines indicate the movements of individual AMRs for ERA-20c and 20CR, respectively. The straight lines represent the linear fit to AMRs from 1900 to 2010.

Table 4. Mann–Kendall test results for the mean of the AMRs of the bias corrected ERA-20c and 20CR by QM approaches from 1900 to 2010 as illustrated in Figure 9. Note that z and b (mm/year) values represent the standardized test statistics and the slope of the trend, respectively, and z values over 1.96 indicate a significant trend at the 0.05 significance level in this test.

Method	ERA-20c		20CR	
	z	b	z	b
gevSQM	5.51	0.50	3.43	0.38
gevQDM	4.21	0.40	2.67	0.34
gamSQM	5.56	0.55	3.53	0.45
gamQDM	4.06	0.41	2.71	0.37
gumSQM	5.50	0.52	3.59	0.42
gumQDM	4.30	0.40	2.91	0.36

It is surprising that the slopes of all stations based on the two different reanalysis products present a similar increasing trend. This result implies that the extreme data over South Korea might have a significant increasing trend that the current observed data with the limited period such as 1974–2010 and 1974–2017 cannot capture. Therefore, the abstraction of these trends might be beneficial in estimating the future extreme design rainfall over South Korea with nonstationary frequency analysis. The following study was conducted accordingly to employ the derived overall trend that might be more feasible to occur in nature, even though a certain degree of uncertainties is included.

4.3. Design Rainfalls with Nonstationary Condition

To explore design rainfalls with the nonstationary condition by using the bias-corrected AMRs from 1900 to 2010, we estimated time-varying parameters of GEV distribution. Nonstationary design rainfalls were estimated by the BM approach based on the GEV parameters ($\mu_{s,m}$, $\mu_{i,m}$, σ_m and ξ_m) derived from the bias-corrected AMRs. Here, the bias-corrected AMRs were collected from QM approaches with GEV distribution (gevSQM and gevQDM), which showed the best performance in Section 4.1.

To find out the impact of nonstationary condition, we representatively illustrated design rainfall comparisons in the selected 4 stations, St.5, St.18, St.21, and St.27 in Figure 10. In the comparisons, design quantiles showed the significant range of uncertainties, which had an upper-bound of approximately 1.3 to 2.0 times higher than the design rainfalls by the observed and a lower bound of about 7–45% lower. For example, for St.5 Seoul, the precipitation quantiles with 100-year return period derived from ERA-20C varied from about 327 mm to 815 mm for gevSQM, and from 346 mm to 852 mm for gevQDM, while the classic quantile by the observation was 483 mm. This characteristic is also shown in the other stations.

For the median values, the design quantiles of the reanalyses with long return periods such as 100-year and 200-year were generally higher than those by the observations except st.21 Busan, while the design rainfalls with short return periods, i.e., 10-year and 20-year, have little difference from the observed. For instance, St.18 Jeonju had a small gap in 10-year design quantiles between the observation and the reanalyses, but the longer the return period, the more gap there exists, especially for ERA-20c. St.5 Seoul and St.27 Inje also showed a similar characteristic, with the design rainfall for the reanalyses exceeding those for the observed as the return period became longer. On the other hand, St.21 Busan had no clear feature compared with the other stations, and even the design rainfall for the observed exceeded those by the bias-corrected reanalyses for gevSQM. Although reanalysis products-based design rainfall estimation includes a certain degree of uncertainty, these results imply that the nonstationary design rainfall would influence on estimating the future risk of extreme precipitation and the strength of the impact depends on the target return period and location.

To find out the spatial influence of non-stationarity in design rainfall, we estimated the relative change (%) of design rainfall with 100-year return period based on the median values of the generated parameter chains in 48 stations. The relative change in 48 stations varied from -38.1% to 58.4% for gevSQM, and from -30.8% to 42.8% for gevQDM, respectively, but the spatial comparisons in Figure 11 illustrated the increase of the relative change (%) in many regions over South Korea.

Furthermore, the spatial area presenting lower or higher than the observed quantiles is more similar in case of the QDM for ERA-20c and 20CR than SQM. The results of the gevQDM relative change shown at the bottom panels of Figure 11 present that the southern and middle regions have higher design rainfall estimation than the observed one, while the northern region and the edge of the southeast and southwest present lower design rainfall from the gevQDM than from the observed one.

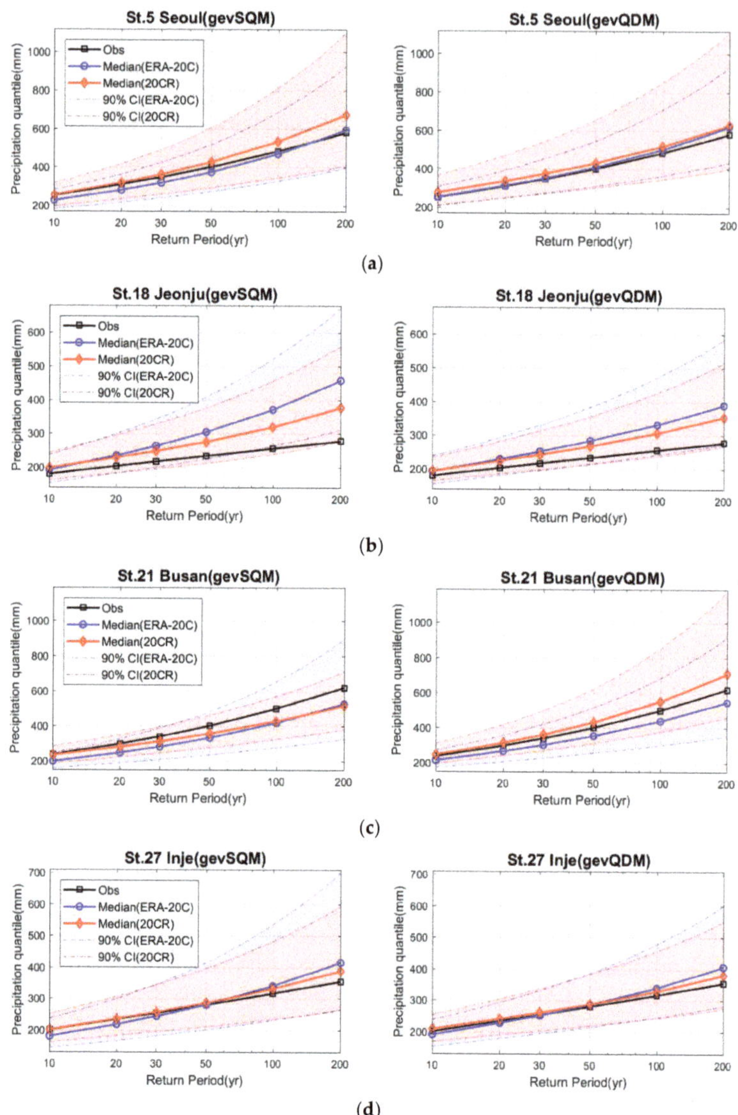

Figure 10. Precipitation quantiles by the conventional GEV model using the AMRs of observation (black line) for the reference period (1974–2010) and the nonstationary GEV models using the bias-corrected ERA-20c and 20CR derived from gevSQM and gevQDM in 5 stations. (**a**) St.5 Seoul, (**b**) St.18 Jeonju, (**c**) St.21 Busan and (**d**) St.27 Inje.

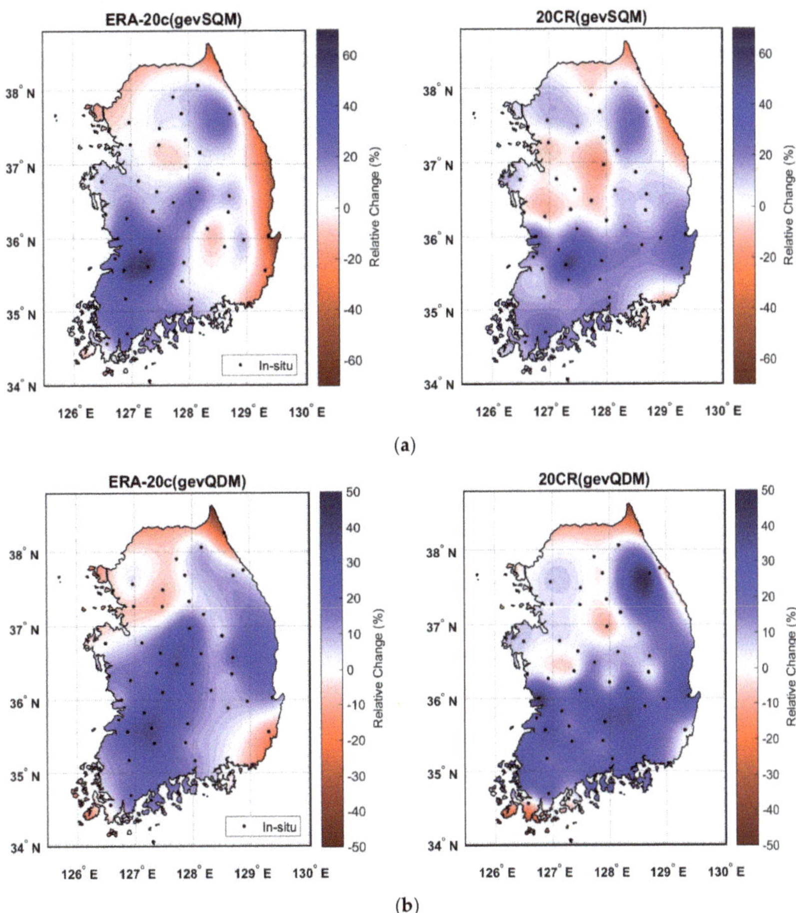

Figure 11. Relative change (%) of design rainfalls with 100-year return period between stationary condition and nonstationary condition. (**a**) indicates the relative change for the bias corrected AMRs by gevSQM, while (**b**) means the results for the bias corrected AMRs by gevQDM.

Despite the uncertainty range, this result suggests that the conventional stationary approach based on the multi-decadal observation may lead to significant underestimation of future risk in some regions. For example, southwest parts within 35–36° N and 126.5–127.5° E had relative change within approximately 10% to 50% in all comparisons. It is obvious that there still exists a certain degree of errors in design rainfall taken from the bias-corrected reanalyses. Nevertheless, if practitioners want to design a project but only have a limited observation, this result can provide meaningful information for a project plan with long-term life span.

5. Discussion

Despite the meaningful information, the proposed method inevitably contains errors from various reasons. Basically, long-term reanalyses for daily precipitation have the systematic errors that spatio-temporally vary [17,20,67–69]. Previous studies have documented that century-long reanalyses like ERA-20c and 20CR are able to mislead long-term trends and the bias may considerably exist for the first half of the twentieth century [16,70–73]. The proposed bias correction methods, SQM and QDM, also have a limitation. As a QM approach cannot exactly correct the climate change trend [23], the potential error in the long-term trend of the raw data may propagate the bias into the bias corrected value. Moreover, a QM approach based on a certain distribution such as gamma can underestimate highly extreme rainfalls, which are mainly described by the upper tail of the distribution [55,74–76]. Thus, some studied have introduced a combination or mixture distribution like gamma-Pareto to better reproduce the heavy tail [24,55,77,78]. The scale gap between the observed and the modelled can result in the biases [79]. In other words, as the proposed QM approaches matched the transfer function between the observation with point-scale and the model data with grid-scale, the bias-corrected value may include errors.

In design rainfall estimation with nonstationary condition, how to define time-varying parameters is also one major issue. We assume a linearly time-dependent location parameter, which is commonly adopted in nonstationary studies, but several studies have also suggested non-linear location parameter or time-varying scale parameter [1,35,38,40,41,43,46,59,66].

Likewise, various causes may result in substantial errors for the design rainfall interpretation in this study. Nevertheless, the proposed analysis suggests meaningful information to help to forecast future risk under the climate change environment. As rainfall intensity is expected to increase in the future, the conventional approach with stationary assumption may underestimate future risk. Several studies and authorities have suggested the guideline for design rainfall and design flood considering the potential impact [80–82]. However, these guidelines typically suggest adding a correction factor into design rainfall or design flood estimated under stationary condition [81]. For instance, the agency [82] recommended increasing the peak rainfall intensity from 5% to 40% for the future period (2015–2115), compared with the data for the baseline period (1961–1990). However, those suggestions are generally based on the analysis of future climate change scenarios which also include significant biases, and the nonstationary analysis based on the observation is constrained by lack of data. Thus, despite the substantial errors, the proposed approach in this study can be a considerable option to achieve additional information for estimating future risk in rainfall.

6. Summary and Conclusions

In this study, we aimed to explore design rainfalls under nonstationary condition by using century-long reanalyses, ERA-20 and 20CR, over South Korea. For this purpose, we first improved the AMRs of the reanalyses from 1900 to 2010 by using a trend-preserving method, QDM, compared with the conventional stationary QM scheme, SQM. After bias correction, we assessed the long-term trend of the bias-corrected AMRs for the whole 20th century to confirm the nonstationarity. With the improved values of ERA-20c and 20CR, we estimated design rainfalls under nonstationary condition based on the ENE approach in 48 stations. Finally, we also explored the spatial change in design rainfalls between the applied nonstationary approach in the current study and the conventional approach. The major results obtained in this study are summarized as follows:

1. The applied QM approaches (gevQM, gamQM and gumQM) significantly improved the AMRs of ERA-20c and 20CR for the reference period. Among three QM schemes, gevQM performed the best in terms of RMSE and NSE.
2. For long-term trend, no significant trend for the AMRs of the observed and the reanalyses can be found during the observational period. However, the century-long AMRs of the bias corrected ERA-20c and 20CR indicated the increasing trends. This result implies that the AMRs might have time-dependent characteristics and the trend in the long-term reanalysis datasets could be beneficial in estimating the future extreme design rainfall over South Korea with nonstationary frequency analysis.
3. The design rainfalls estimated under nonstationary condition were influenced in estimating the future risk of extreme precipitation and the strength of the impact depends on the target return period and location. More specifically, the nonstationary design rainfalls in some parts of South Korea exceeded the classic design rainfalls by the observed. This result implies that the nonstationarity in the AMRs that the short-

term observation often fails to detect could deteriorate the confidence of a project based on the observed data for the future risk in South Korea.

The findings obtained in this study provide a meaningful perspective on the applicability of century-long reanalysis products, especially for nonstationary rainfall frequency analysis in a region with a limited observation network. Despite a certain degree of errors, the proposed scheme with employing the reanalysis products can be beneficial to predict the future evolution of extreme precipitation and to estimate the design rainfall accordingly.

Supplementary Materials: The following are available online at https://www.mdpi.com/2073-4433/12/2/191/s1. Figure S1. A flow chart for estimating design rainfall under the conditions of nonstationarity and stationarity.

Author Contributions: Methodology, D.-I.K. and T.L.; Project administration, D.H.; Writing—original draft, D.-I.K.; Writing—review & editing, T.L. All authors have read and agreed to the published version of the manuscript.

Funding: The first author is funded by the Government of South Korea for performing his doctoral studies at the University of Bristol. The third author acknowledges the support of the National Research Foundation of Korea (NRF) grant (2018R1A2B600179), funded by the Korean Government (MEST).

Institutional Review Board Statement: Not applicable.

Informed Consent Statement: Not applicable.

Data Availability Statement: Not applicable.

Acknowledgments: We are grateful for the relevant data provided by KMA, ECMWF and NCAA.

Conflicts of Interest: The authors declare no conflict of interest.

References

1. Cheng, L.; Aghakouchak, A. Nonstationary precipitation intensity-duration-frequency curves for infrastructure design in a changing climate. *Sci. Rep.* **2014**, *4*, 1–6. [CrossRef] [PubMed]
2. Li, J.; Evans, J.; Johnson, F.; Sharma, A. A comparison of methods for estimating climate change impact on design rainfall using a high-resolution RCM. *J. Hydrol.* **2017**, *547*, 413–427. [CrossRef]
3. Alexander, L.V.; Zhang, X.; Peterson, T.C.; Caesar, J.; Gleason, B.; Klein Tank, A.M.G.; Haylock, M.; Collins, D.; Trewin, B.; Rahimzadeh, F.; et al. Global observed changes in daily climate extremes of temperature and precipitation. *J. Geophys. Res. Atmos.* **2006**, *111*, 1–22. [CrossRef]
4. IPCC. *Climate Change 2014–Impacts, Adaptation and Vulnerability: Regional Aspects*; Cambridge University Press: London, UK, 2014.
5. De Luca, D.L.; Galasso, L. Stationary and non-stationary frameworks for extreme rainfall time series in southern Italy. *Water* **2018**, *10*, 1477. [CrossRef]
6. Coles, S.; Pericchi, L.R.; Sisson, S. A fully probabilistic approach to extreme rainfall modeling. *J. Hydrol.* **2003**, *273*, 35–50. [CrossRef]
7. Overeem, A.; Buishand, A.; Holleman, I. Rainfall depth-duration-frequency curves and their uncertainties. *J. Hydrol.* **2008**, *348*, 124–134. [CrossRef]
8. Huard, D.; Mailhot, A.; Duchesne, S. Bayesian estimation of intensity-duration-frequency curves and of the return period associated to a given rainfall event. *Stoch. Environ. Res. Risk Assess.* **2010**, *24*, 337–347. [CrossRef]
9. Tung, Y.K.; Wong, C.L. Assessment of design rainfall uncertainty for hydrologic engineering applications in Hong Kong. *Stoch. Environ. Res. Risk Assess.* **2014**, *28*, 583–592. [CrossRef]
10. Van de Vyver, H. Bayesian estimation of rainfall intensity-duration-frequency relationships. *J. Hydrol.* **2015**, *529*, 1451–1463. [CrossRef]
11. Serinaldi, F.; Kilsby, C.G. Stationarity is undead: Uncertainty dominates the distribution of extremes. *Adv. Water Resour.* **2015**, *77*, 17–36. [CrossRef]
12. Koutsoyiannis, D.; Montanari, A. Negligent killing of scientific concepts: The stationarity case. *Hydrol. Sci. J.* **2015**, *60*, 1174–1183. [CrossRef]
13. Dee, D.P.; Uppala, S.M.; Simmons, A.J.; Berrisford, P.; Poli, P.; Kobayashi, S.; Andrae, U.; Balmaseda, M.A.; Balsamo, G.; Bauer, P. The ERA-Interim reanalysis: Configuration and performance of the data assimilation system. *Q. J. R. Meteorol. Soc.* **2011**, *137*, 553–597. [CrossRef]
14. Zhang, Q.; Körnich, H.; Holmgren, K. How well do reanalyses represent the southern African precipitation? *Clim. Dyn.* **2013**, *40*, 951–962. [CrossRef]

15. Hersbach, H.; Peubey, C.; Simmons, A.; Berrisford, P.; Poli, P.; Dee, D. ERA-20CM: A twentieth-century atmospheric model ensemble. *Q. J. R. Meteorol. Soc.* **2015**, *141*, 2350–2375. [CrossRef]
16. Donat, M.G.; Alexander, L.V.; Herold, N.; Dittus, A.J. Temperature and precipitation extremes in century-long gridded observations, reanalyses, and atmospheric model simulations. *J. Geophys. Res. Atmos.* **2016**, *121*, 11174–11189. [CrossRef]
17. Gao, L.; Bernhardt, M.; Schulz, K.; Chen, X.W.; Chen, Y.; Liu, M.B. A First Evaluation of ERA-20CM over China. *Mon. Weather Rev.* **2016**, *144*, 45–57. [CrossRef]
18. Poli, P.; Hersbach, H.; Dee, D.P.; Berrisford, P.; Simmons, A.J.; Vitart, F.; Laloyaux, P.; Tan, D.G.H.; Peubey, C.; Thépaut, J.-N. ERA-20C: An Atmospheric Reanalysis of the Twentieth Century. *J. Clim.* **2016**, *29*, 4083–4097. [CrossRef]
19. Compo, G.P.; Whitaker, J.S.; Sardeshmukh, P.D.; Matsui, N.; Allan, R.J.; Yin, X.; Gleason, B.E.; Vose, R.S.; Rutledge, G.; Bessemoulin, P. The twentieth century reanalysis project. *Q. J. R. Meteorol. Soc.* **2011**, *137*, 1–28. [CrossRef]
20. Kim, D.-I.; Han, D. Comparative study on long term climate data sources over South Korea. *J. Water Clim. Chang.* **2018**, (in press). [CrossRef]
21. Teutschbein, C.; Seibert, J. Bias correction of regional climate model simulations for hydrological climate-change impact studies: Review and evaluation of different methods. *J. Hydrol.* **2012**, *456*, 12–29. [CrossRef]
22. Mao, G.; Vogl, S.; Laux, P.; Wagner, S.; Kunstmann, H. Stochastic bias correction of dynamically downscaled precipitation fields for Germany through Copula-based integration of gridded observation data. *Hydrol. Earth Syst. Sci.* **2015**, *19*, 1787–1806. [CrossRef]
23. Maraun, D. Bias Correcting Climate Change Simulations—A Critical Review. *Curr. Clim. Chang. Rep.* **2016**, *2*, 211–220. [CrossRef]
24. Nyunt, C.T.; Koike, T.; Yamamoto, A. Statistical bias correction for climate change impact on the basin scale precipitation in Sri Lanka, Philippines, Japan and Tunisia. *Hydrol. Earth Syst. Sci.* **2016**. [CrossRef]
25. Themeßl, M.J.; Gobiet, A.; Heinrich, G. Empirical-statistical downscaling and error correction of regional climate models and its impact on the climate change signal. *Clim. Chang.* **2012**, *112*, 449–468. [CrossRef]
26. Fang, G.; Yang, J.; Chen, Y.N.; Zammit, C. Comparing bias correction methods in downscaling meteorological variables for a hydrologic impact study in an arid area in China. *Hydrol. Earth Syst. Sci.* **2015**, *19*, 2547–2559. [CrossRef]
27. Maraun, D.; Widmann, M. *Statistical Downscaling and Bias Correction for Climate Research*; Cambridge University Press: London, UK, 2018; pp. 170–200.
28. Chang, H.; Kwon, W.-T. Spatial variations of summer precipitation trends in South Korea, 1973–2005. *Environ. Res. Lett.* **2007**, *2*, 45012. [CrossRef]
29. Choi, G.; Collins, D.; Ren, G.; Trewin, B.; Baldi, M.; Fukuda, Y.; Afzaal, M.; Pianmana, T.; Gomboluudev, P.; Huong, P.T.T. Changes in means and extreme events of temperature and precipitation in the Asia-Pacific Network region, 1955–2007. *Int. J. Climatol.* **2009**, *29*, 1906–1925. [CrossRef]
30. Jung, I.W.; Bae, D.H.; Kim, G. Recent trends of mean and extreme precipitation in Korea. *Int. J. Climatol.* **2011**, *31*, 359–370. [CrossRef]
31. Cannon, A.J.; Sobie, S.R.; Murdock, T.Q. Bias correction of GCM precipitation by quantile mapping: How well do methods preserve changes in quantiles and extremes? *J. Clim.* **2015**, *28*, 6938–6959. [CrossRef]
32. Miao, C.; Su, L.; Sun, Q.; Duan, Q. A nonstationary bias-correction technique to remove bias in GCM simulations. *J. Geophys. Res.* **2016**, *121*, 5718–5735. [CrossRef]
33. Eum, H.I.; Cannon, A.J. Intercomparison of projected changes in climate extremes for South Korea: Application of trend preserving statistical downscaling methods to the CMIP5 ensemble. *Int. J. Climatol.* **2017**, *37*, 3381–3397. [CrossRef]
34. Nahar, J.; Johnson, F.; Sharma, A. Assessing the extent of non-stationary biases in GCMs. *J. Hydrol.* **2017**, *549*, 148–162. [CrossRef]
35. Nadarajah, S.; Choi, D. Maximum daily rainfall in South Korea. *J. Earth Syst. Sci.* **2007**, *116*, 311–320. [CrossRef]
36. Cunderlik, J.M.; Burn, D.H. Non-stationary pooled flood frequency analysis. *J. Hydrol.* **2003**, *276*, 210–223. [CrossRef]
37. El Adlouni, S.; Ouarda, T.B.M.J.; Zhang, X.; Roy, R.; Bobée, B. Generalized maximum likelihood estimators for the nonstationary generalized extreme value model. *Water Resour. Res.* **2007**, *43*, 1–13. [CrossRef]
38. Leclerc, M.; Ouarda, T.B.M.J. Non-stationary regional flood frequency analysis at ungauged sites. *J. Hydrol.* **2007**, *343*, 254–265. [CrossRef]
39. Cannon, A.J. A flexible nonlinear modelling framework for nonstationary generalized extreme value analysis in hydroclimatology. *Hydrol. Process.* **2010**, *24*, 673–685. [CrossRef]
40. Panagoulia, D.; Economou, P.; Caroni, C. Stationary and nonstationary generalized extreme value modelling of extreme precipitation over a mountainous area under climate change. *Environmetrics* **2014**, *25*, 29–43. [CrossRef]
41. Son, C.; Lee, T.; Kwon, H.H. Integrating nonstationary behaviors of typhoon and non-typhoon extreme rainfall events in East Asia. *Sci. Rep.* **2017**, *7*, 1–9. [CrossRef]
42. Du, T.; Xiong, L.; Xu, C.Y.; Gippel, C.J.; Guo, S.; Liu, P. Return period and risk analysis of nonstationary low-flow series under climate change. *J. Hydrol.* **2015**, *527*, 234–250. [CrossRef]
43. Cheng, L.; AghaKouchak, A.; Gilleland, E.; Katz, R.W. Non-stationary extreme value analysis in a changing climate. *Clim. Chang.* **2014**, *127*, 353–369. [CrossRef]
44. Salas, J.D.; Obeysekera, J. Revisiting the Concepts of Return Period and Risk for Nonstationary Hydrologic Extreme Events. *J. Hydrol. Eng.* **2014**, *19*, 554–568. [CrossRef]
45. Read, L.K.; Vogel, R.M. Reliability, return periods, and risk under nonstationarity. *Water Resour. Res.* **2015**, *51*, 6381–6398. [CrossRef]

46. Obeysekera, J.; Salas, J.D. Frequency of Recurrent Extremes under Nonstationarity. *J. Hydrol. Eng.* **2016**, *21*, 1–9. [CrossRef]
47. Salas, J.D.; Obeysekera, J.; Vogel, R.M. Techniques for assessing water infrastructure for nonstationary extreme events: A review. *Hydrol. Sci. J.* **2018**, *63*, 325–352. [CrossRef]
48. Berntell, E.; Zhang, Q.; Chafik, L.; Körnich, H. Representation of Multidecadal Sahel Rainfall Variability in 20th Century Reanalyses. *Sci. Rep.* **2018**, *8*. [CrossRef]
49. Diro, G.T.; Grimes, D.I.F.; Black, E.; O'Neill, A.; Pardo-Iguzquiza, E. Evaluation of reanalysis rainfall estimates over Ethiopia. *Int. J. Climatol.* **2009**, *29*, 67–78. [CrossRef]
50. Hu, Q.; Li, Z.; Wang, L.; Huang, Y.; Wang, Y.; Li, L. Rainfall spatial estimations: A review from spatial interpolation to multi-source data merging. *Water* **2019**, *11*, 579. [CrossRef]
51. Hua, W.; Zhou, L.; Nicholson, S.E.; Chen, H.; Qin, M. Assessing reanalysis data for understanding rainfall climatology and variability over Central Equatorial Africa. *Clim. Dyn.* **2019**, *53*, 651–669. [CrossRef]
52. Kim, K.B.; Bray, M.; Han, D. An improved bias correction scheme based on comparative precipitation characteristics. *Hydrol. Process.* **2015**, *29*, 2258–2266. [CrossRef]
53. Rabiei, E.; Haberlandt, U. Applying bias correction for merging rain gauge and radar data. *J. Hydrol.* **2015**, *522*, 544–557. [CrossRef]
54. Kim, K.B.; Kwon, H.H.; Han, D. Bias correction methods for regional climate model simulations considering the distributional parametric uncertainty underlying the observations. *J. Hydrol.* **2015**, *530*, 568–579. [CrossRef]
55. Volosciuk, C.; Maraun, D.; Vrac, M.; Widmann, M. A combined statistical bias correction and stochastic downscaling method for precipitation. *Hydrol. Earth Syst. Sci.* **2017**, *21*, 1693–1719. [CrossRef]
56. Koutsoyiannis, D. Statistics of extremes and estimation of extreme rainfall: I. Theoretical investigation. *Hydrol. Sci. J.* **2004**, *49*, 575–590. [CrossRef]
57. Wilson, P.S.; Toumi, R. A fundamental probability distribution for heavy rainfall. *Geophys. Res. Lett.* **2005**, *32*, 1–4. [CrossRef]
58. Li, H.; Sheffield, J.; Wood, E.F. Bias correction of monthly precipitation and temperature fields from Intergovernmental Panel on Climate Change AR4 models using equidistant quantile matching. *J. Geophys. Res. Atmos.* **2010**, *115*, 1–20. [CrossRef]
59. Bürger, G.; Sobie, S.R.; Cannon, A.J.; Werner, A.T.; Murdock, T.Q. Downscaling extremes: An intercomparison of multiple methods for future climate. *J. Clim.* **2013**, *26*, 3429–3449. [CrossRef]
60. Hamed, K.H.; Rao, A.R. A modified Mann-Kendall trend test for autocorrelated data. *J. Hydrol.* **1998**, *204*, 182–196. [CrossRef]
61. Sen, P.K. Estimates of the regression coefficient based on Kendall's tau. *J. Am. Stat. Assoc.* **1968**, *63*, 1379–1389. [CrossRef]
62. Theil, H. A rank-invariant method of linear and polynomial regression analysis. I, II, III. *Proc. K. Ned. Akad. Wet.* **1950**, *53*, 386–392, 521–525, 1397–1412.
63. Sayemuzzaman, M.; Jha, M.K. Seasonal and annual precipitation time series trend analysis in North Carolina, United States. *Atmos. Res.* **2014**, *137*, 183–194. [CrossRef]
64. Shadmani, M.; Marofi, S.; Roknian, M. Trend analysis in reference evapotranspiration using Mann-Kendall and Spearman's Rho tests in arid regions of Iran. *Water Resour. Manag.* **2012**, *26*, 211–224. [CrossRef]
65. Ouarda, T.B.M.J.; El-Adlouni, S. Bayesian nonstationary frequency analysis of hydrological variables. *J. Am. Water Resour. Assoc.* **2011**, *47*, 496–505. [CrossRef]
66. Obeysekera, J.; Salas, J.D. Quantifying the Uncertainty of Design Floods under Nonstationary Conditions. *J. Hydrol. Eng.* **2014**, *19*, 1438–1446. [CrossRef]
67. Bosilovich, M.G.; Chen, J.; Robertson, F.R.; Adler, R.F. Evaluation of global precipitation in reanalyses. *J. Appl. Meteorol. Climatol.* **2008**, *47*, 2279–2299. [CrossRef]
68. Ma, L.; Zhang, T.; Frauenfeld, O.W.; Ye, B.; Yang, D.; Qin, D. Evaluation of precipitation from the ERA-40, NCEP-1, and NCEP-2 Reanalyses and CMAP-1, CMAP-2, and GPCP-2 with ground-based measurements in China. *J. Geophys. Res. Atmos.* **2009**, *114*, 1–20. [CrossRef]
69. Bao, X.; Zhang, F. Evaluation of NCEP–CFSR, NCEP–NCAR, ERA-Interim, and ERA-40 reanalysis datasets against independent sounding observations over the Tibetan Plateau. *J. Clim.* **2013**, *26*, 206–214. [CrossRef]
70. Brands, S.; Gutiérrez, J.M.; Herrera, S.; Cofiño, A.S. On the use of reanalysis data for downscaling. *J. Clim.* **2012**, *25*, 2517–2526. [CrossRef]
71. Krueger, O.; Schenk, F.; Feser, F.; Weisse, R. Inconsistencies between long-term trends in storminess derived from the 20CR reanalysis and observations. *J. Clim.* **2013**, *26*, 868–874. [CrossRef]
72. Poli, P.; Hersbach, H.; Tan, D.; Dee, D.; Thépaut, J.-N.; Simmons, A.; Peubey, C.; Laloyaux, P.; Komori, T.; Berrisford, P.; et al. The data assimilation system and initial performance evaluation of the ECMWF pilot reanalysis of the 20th-century assimilating surface observations only (ERA-20C). *ERA Rep. Ser.* **2013**, *14*, 59.
73. Befort, D.J.; Wild, S.; Kruschke, T.; Ulbrich, U.; Leckebusch, G.C. Different long-term trends of extra-tropical cyclones and windstorms in ERA-20C and NOAA-20CR reanalyses. *Atmos. Sci. Lett.* **2016**, *17*, 586–595. [CrossRef]
74. Wilks, D.S. Interannual variability and extreme-value characteristics of several stochastic daily precipitation models. *Agric. For. Meteorol.* **1999**, *93*, 153–169. [CrossRef]
75. Vrac, M.; Naveau, P. Stochastic downscaling of precipitation: From dry events to heavy rainfalls. *Water Resour. Res.* **2007**, *43*, 1–13. [CrossRef]

76. Hundecha, Y.; Pahlow, M.; Schumann, A. Modeling of daily precipitation at multiple locations using a mixture of distributions to characterize the extremes. *Water Resour. Res.* **2009**, *45*, 1–15. [CrossRef]
77. Gutjahr, O.; Heinemann, G. Comparing precipitation bias correction methods for high-resolution regional climate simulations using COSMO-CLM. *Theor. Appl. Climatol.* **2013**, *114*, 511–529. [CrossRef]
78. Smith, A.; Freer, J.; Bates, P.; Sampson, C. Comparing ensemble projections of flooding against flood estimation by continuous simulation. *J. Hydrol.* **2014**, *511*, 205–219. [CrossRef]
79. Maraun, D. Bias Correction, Quantile Mapping, and Downscaling: Revisiting the Inflation Issue. *J. Clim.* **2013**, *26*, 2013–2014. [CrossRef]
80. Lawrence, D.; Hisdal, H. *Hydrological Projections for Floods in Norway under a Future Climate*; Norwegian Water Resources and Energy Directorate: Oslo, Norway, 2011; p. 47.
81. Madsen, H.; Lawrence, D.; Lang, M.; Martinkova, M.; Kjeldsen, T.R. Review of trend analysis and climate change projections of extreme precipitation and floods in Europe. *J. Hydrol.* **2014**, *519*, 3634–3650. [CrossRef]
82. Flood Risk Assessments: Climate Change Allowances. 2017. Available online: https://www.gov.uk/guidance/flood-risk-assessments-climate-change-allowances (accessed on 31 January 2020).

Article

Changes in Intensity and Variability of Tropical Cyclones over the Western North Pacific and Their Local Impacts under Different Types of El Niños

Yuhang Liu [1], Sun-Kwon Yoon [2,*], Jong-Suk Kim [1,*], Lihua Xiong [1] and Joo-Heon Lee [3]

1 State Key Laboratory of Water Resources and Hydropower Engineering Science, Wuhan University, Wuhan 430072, China; yuhangliu233@gmail.com (Y.L.); xionglh@whu.edu.cn (L.X.)
2 Department of Safety and Disaster Prevention Research, Seoul Institute of Technology, Seoul 03909, Korea
3 Department of Civil Engineering, Joongbu University, Goyang-si, Gyeunggi-do 10279, Korea; leejh@joongbu.ac.kr
* Correspondence: skyoon@sit.re.kr (S.-K.Y.); jongsuk@whu.edu.cn (J.-S.K.)

Abstract: This study investigated the effects of El Niño events on tropical cyclone (TC) characteristics over the western North Pacific (WNP) region. First, TC characteristics associated with large-scale atmospheric phenomena (i.e., genesis position, frequency, track, intensity, and duration) were investigated in the WNP in relation to various types of El Niño events—moderate central Pacific (MCP), moderate eastern Pacific (MEP), and strong basin-wide (SBW). Subsequently, the seasonal and regional variability of TC-induced rainfall across China was analyzed to compare precipitation patterns under the three El Niño types. When extreme El Niño events of varying degrees occurred, the local rainfall varied during the developmental and decaying years. The development of MEP and SBW was associated with a distinct change in TC-induced rainfall. During MEP development, TC-induced rainfall occurred in eastern and northeastern China, whereas in SBW, TC-induced heavy rainfall occurred in southwest China. During SBW development, the southwestern region was affected by TCs over a long period, with the eastern and northeastern regions being affected significantly fewer days. During El Niño decay, coastal areas were relatively more affected by TCs during MCP events, and the Pearl River basin was more affected during SBW events. This study's results could help mitigate TC-related disasters and improve water-supply management.

Keywords: extreme El Niño event; tropical cyclone; tropical cyclone-induced precipitation; China

Citation: Liu, Y.; Yoon, S.-K.; Kim, J.-S.; Xiong, L.; Lee, J.-H. Changes in Intensity and Variability of Tropical Cyclones over the Western North Pacific and Their Local Impacts under Different Types of El Niños. *Atmosphere* **2021**, *12*, 59. https://doi.org/10.3390/atmos12010059

Received: 12 November 2020
Accepted: 29 December 2020
Published: 31 December 2020

1. Introduction

Recent advances in satellite remote sensing have provided the opportunity to assess the impact of anthropogenic climate change on natural disasters, a topic that has been explored in many studies [1–4]. Data obtained in studies on climate change can provide the data basis for environmental monitoring and forecasting, which have implications for global agriculture, livestock breeding, forestry, and other natural-resource industries. Damage to industries and infrastructure from events, such as floods or droughts, could be mitigated through an improved understanding of the effects of climate change; for example, knowledge of flood trends could support dam flood-control calculations and repairs [1]. Many studies have shown that tropical cyclones (TCs) have a considerable impact on China's summer rainfall. Studying TC patterns could help link summer hydrological decision-making in China with global climate trends. In the western North Pacific (WNP), TC-induced rain accounted for 9% of rainfall events and 21% of accumulated rainfall for 1979–2005 [2]. TC events accounted for 37% of natural disaster-related damage in China, and from 1994 to 2013, catastrophic TC events led to an economic loss of approximately 17% in China [3]. Therefore, it is necessary to investigate the changing intensity and variability of TCs, as well as their local impacts.

TC-induced rainfall is affected by many factors, including the El Niño–Southern Oscillation (ENSO) [4,5]. Air–sea interactions in the tropical Pacific generate extreme ENSO events [5]. Zhao et al. [3] investigated TC variability using satellite data from 1979 to 2015, and along with a decreasing trend in the number of TCs in the WNP, they observed a stronger relationship between El Niño and TC activities since 1998. Thus, it is vital to better understand the relationship between extreme ENSO events and TC characteristics to improve climate projections [4]. Many studies have explored factors influencing seasonal precipitation variations, and El Niño events have been found to greatly impact summer precipitation in China [4,6–8]. Additionally, TCs that occur in the WNP promote extreme rainfall events accompanied by abnormal advection of moisture. However, El Niño's effect on rainfall and its indirect impact on the behavior of TCs remain partially understood [4,9]. Therefore, in this study, we evaluated the impact of different types of El Niño events on TC activities and TC-induced rainfall.

A subtropical high is one of several highs that reach the Northern Hemisphere and play an important role in balancing transportation, water supply, energy, and heat, as well as effecting seasonal variations. Thus, the subtropical high has a strong influence on TC tracks and the atmospheric climate of China. Many studies have discussed the linkages between TCs and the subtropical high and their role in increasing rainfall. TCs and TC-induced anticyclonic circulation anomalies promote water vapor transport [10–12]. TC events also have a significant impact on the subtropical high and relevant weather changes in the WNP [11,12]. These TC/subtropical high interactions affect the prediction of TC features, such as tracks and heavy rainfall, across East Asia, where the western subtropical high greatly affects the summer climate in East Asia [13].

An investigation of the effects of El Niño on precipitation revealed that ENSO events cause an intense southwest wind along coastal China, which leads to unusual rainfall in northern China in the decaying years of El Niño events [6]. Zhang et al. [7,8] noted that the amounts of TC-induced precipitation in inland China were low compared with that experienced in coastal regions. Further, several studies have indicated that ENSO events affect TC characteristics, such as their tracks and genesis positions [9,14], and their intensity [15,16]. In addition, TCs tend to occur more frequently and have a longer lifetime when occurring in the context of a strong ENSO [17]. However, few studies have emphasized the distinctions between types of El Niño events and their effect on TCs. As the characteristics of TCs and impacts induced by TCs vary, they must be studied separately. From a long-term perspective, discussing links between El Niño events and the impacts of TCs, such as induced rainfall, is crucial.

Zonal and meridional asymmetries in weather and oceanic conditions contribute to the diversity of ENSO forms, which reflect the complicacy of atmospheric feedbacks [18]. Some studies categorized El Niño events as Central Pacific (CP) or Eastern Pacific (EP) events. Cai et al. [19] noted that during the development of EP El Niño events, seasonal rainfall decreases in South America, whereas it increases during El Niño decay. However, in CP events, precipitation patterns are more complicated and remain unclear. Wang et al. [20] applied a nonlinear K-means cluster analysis to distinguish El Niño events of various magnitudes with greater specificity. The method classifies El Niño events based on three types of onsets: moderate central Pacific (MCP), moderate eastern Pacific (MEP), and strong basin-wide (SBW) events. SBW El Niño events are significant in their intensity, particularly in the initial winter and spring, with maximum SSTAs reaching over 2.5 °C [21]. They are coupled with strong atmospheric convection and warm SSTAs, and experience extraordinary basin-wide growth in the boreal spring. They induce SSTAs that spread easterly, causing abnormalities in the west. MCP El Niño events originate in the CP with continuous moderate warming. They extend eastward and reach their maximum intensity in the CP, accompanied by convective wind anomalies. In contrast to SBW activities, the western SSTA in MCP events occurs in the interior of the basin and develops later. MEP El Niño events begin in the far EP. Owing to SSTA-induced negative wind-pressure anomalies that facilitate the westward movement of warming by inhibiting upwelling, the easterly wind anomalies are uniquely and significantly reversed in the central-western Pacific [20].

The amount of rain varies in the emerging and warming phases of El Niño, and the spatiotemporal changes in water vapor tracks depend on the El Niño types [21]. As the ENSO classification system becomes increasingly refined, the relationship between El Niño and TCs characteristics can be analyzed in greater detail. Owing to the delayed atmospheric response to oceanic behavior in El Niño, better knowledge of regional precipitation patterns could simplify rain forecasting [22]. Therefore, this study analyzed differences between the effects of three El Niño types (MCP, MEP, and SBW) on TC activities in the WNP and on the seasonal variability of TC-induced rainfall in China. In addition, this study analyzed rainfall indices for fractional TC-induced precipitation during the study period and applied them to determine optimum forecasting models for the study region. The results of this study could support measures to reduce and mitigate TC-related disasters and improve water-supply management under changing climate conditions.

2. Materials and Methods

2.1. Study Area and Weather Stations

Figure 1 shows the study area and location of active weather stations in China. TC impacts are experienced throughout China, but especially in the country's southeastern area and the Pearl River basin. Therefore, this study selected the area spanning from the equator (EQ) to 60° N and 60° E to 180° E, which includes all regions of China and the areas tracked by TCs. When analyzing the tracks of TCs throughout Asia, the application of a grid map allows the effective visualization of the density of TC distribution. This study utilized daily precipitation data from 839 meteorological stations within nine river basins in China, which were provided by China's Meteorological Data Service Center. Some relatively early data were missing because of observation errors and underdeveloped equipment. Therefore, we selected daily accumulated rainfall data from 1961 to 2017 at 839 stations to measure the influence of different TC patterns on all river basins throughout China.

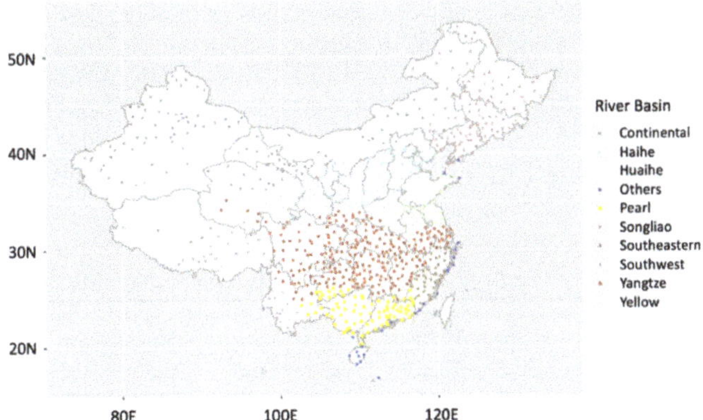

Figure 1. Map of the study area (5–55° N, 75–135° E) and locations of the weather stations in China.

To study the local impacts of TCs, this study considered TC-induced precipitation separately from daily rainfall, which increased the accuracy of the comparative analysis. Kim et al. [23] used different time windows and TC radii to consider the spatiotemporal impacts of TCs. Time windows, an empirical methodology developed by Kim and Jain [23] to study seasonal flow separation in the Korean Peninsula, specify the timeframes impacted by TCs and extract TC-induced rainfall from daily precipitation data. This study adopted a three-day time window to reflect the duration of the relevant moisture supply [24]. Among precipitation indices widely applied in previous studies to perform trend analyses and regional rainfall calculations [7,8,25], this study selected seasonal total precipitation on wet days when daily precipitation >1 mm (PRCPTOT), the maximum number of consecutive TC-induced precipitation days (CTPD), to help to analyze regional rainfall characteristics and summarize the distribution of precipitation.

2.2. Classification of El Niño Events

To distinguish the impacts of diverse TCs, this study focused on TCs that occurred during different types of El Niño events. By applying the K-means cluster analysis to the development of the SSTA for 33 extreme El Niño events from 1901 to 2017, first-year El Niño events were classified into three types—MCP, MEP, and SBW—which vary in origin, initiation, outspreading, strength, and other features (see Table 1). These parameters were determined based on the Niño 3.4 index, which is derived from SSTAs calculated using the Hadley Center Sea Ice and SST dataset version 1 (HadISST1) from the Met Office Marine Data Bank and WMO GTS data by removing the mean SST in the region ($170°$ W–$120°$ W, $5°$ S–$5°$ N). "Extreme" El Niño events were defined according to Wang et al. [20]. The SST data and Niño 3.4 index values were obtained from the Ocean Observations Panel for Climate (OOPC) and the Joint WMO-IOC Commission for Oceanography and Marine Meteorology (JCOMM) Services program area. They served to verify the characteristics of selected El Niño events to reduce errors owing to different data-capture regions.

Table 1. Main properties of moderate central Pacific (MCP), moderate eastern Pacific (MEP), and strong basin-wide (SBW) ENSO events.

Moderate Central Pacific (MCP)	Moderate Eastern Pacific (MEP)	Strong Basin-Wide (SBW)
• begin with mild warming in the western Pacific • expand eastward	• originate in the far EP • propagate westward	• westerly SST anomaly followed by distinctive basin-wide development • extraordinary intensity (maximum SSTA >2.5 °C)

In general, during the mature phases of MCP, MEP, and SBW events, their associated SSTAs are clearly distinguishable in terms of temperature distribution and intensities and thus result in spatiotemporal precipitation and track variations [20]. Considering their intensity and rapid propagation, the developing years are defined as the years in which El Niño occurs. Our cluster-based approach classified 1986/87, 1991/92, 1994/95, 2002/03, 2004/05, 2006/07, 2009/10, and 2014/15 as years of MCP; 1957/58, 1963/64, 1965/66, 1968/69, and 1976/77 as years of MEP; and 1972/73, 1982/83, 1997/98, and 2015/16 as years of SBW.

2.3. Tropical Cyclone Data

The Regional Specialized Meteorological Center (RSMC), a Tokyo-based meteorological organization, provided the original data and attributes for TCs. These data describe the propagation of TCs, including the geographical coordinates of the TC center, the central pressure, the TC identification number at fixed 6-h time intervals from 1951 to 2017. This information is also recorded in the dataset describing the strength of the max wind speed since 1977. Among these data, this study focused on those pertinent to TCs that occurred in the selected decaying years and affected China, including their genesis position, tracks recorded at 6-h intervals, recurving location (location at which the TC, at its westernmost position, shifts its track to the northeast), central pressure, wind speed, and duration. In general, non-recurving TCs were excluded when calculating the recurving location density over an area [26]. Non-recurving TCs represented 72.04% (for MCP events), 78.57% (for MEP events), and 83.72% (for SBW events) of events. The subtropical high was another considered parameter. The subtropical high (based on the contour line of 5880 gpm at 500 hPa) was derived from the National Centers for Environmental Prediction/National Center for Atmospheric Research (NCEP/NCAR) reanalysis 1 [27].

TC tracks are defined as a series of points over which the TC centers (defined as the center of a circle) pass. Therefore, it is important to define a suitable TC radius to distinguish TC-induced effects from other weather patterns. Assuming all TCs have an equal effect over China, a fixed radius of 500 km was found to include some non-TC-induced rainfall in the total calculated [28]. Mature TC winds usually impact a radius of less than 500 km [26]. Thus, to obtain more accurate statistics on TC-induced precipitation, this study considered a range of TC radii (400–600 km) to reduce error related to non-TC-induced rainfall. This study extracted TC-induced rainfall based on these tracks from the total rainfall during May to October in the selected years using an n-day time window. On days of TC landfall, most rainfall could be assumed to be TC-induced; thus, all rainfall from these days was included.

2.4. Methodology

2.4.1. Kriging

In statistical calculations, primarily in geostatistics, kriging (or Gaussian process regression) is an interpolation method modeled on Gaussian procedures wherein interpolated values are controlled by prior covariances, thus providing a computationally stable and efficient output [28]. Kriging is often applied in the fields of geographic science, environmental science, and atmospheric science [29]. Meteorological data, such as precipitation, have significant spatial characteristics. Since this kind of data comes from spatially discretely distributed monitoring stations, which are unevenly distributed and limited in number, the kriging method is applied to make an overall and coherent rainfall distribution. Many studies have used this kriging method for distribution analysis [30–32].

2.4.2. Welch's *t*-Test

The *t*-test is an inference-based statistical method used to determine whether differences exist between different samples [33]. Among these, Welch's *t*-test is applied when the number of items within each set is different and the variances of the two groups are unknown or unidentical to identify discrepancies. Welch's *t*-test was selected to compare rainfall indices during different El Niño events with a long-term mode, which have usually been considered "unpaired" or "independent samples" [34].

Decadal hydro-meteorological indices, with a relatively small and nonstationary number of samples, meet the requirements for Welch's *t*-test [35]. Welch's *t*-test is defined as follows by the statistic t in the following formula (Equation (1)):

$$t = \frac{\overline{X}_1 - \overline{X}_2}{\sqrt{\frac{S_1^2}{n_1} + \frac{S_2^2}{n_2}}} \tag{1}$$

where $\overline{X}_1$, $\overline{X}_2$ are the sample means, n_1 and n_2 are the sample sizes for sample 1 and sample 2, respectively. S_1^2 and S_1^2 are the sample variances.

The variance is a function calculated as follows (Equation (2)):

$$S^2 = \frac{\sum \left(X - \overline{X} \right)^2}{n - 1} \tag{2}$$

By selecting data that satisfied t at the 95% significance level, this study ensured that the included data were representative of the stations in each river basin and were sensitive to TCs.

2.4.3. Pettitt Test

The Pettitt test is a nonparametric test used to assess the occurrence of sudden changes in records [36]. The Pettitt test is the most commonly utilized method of breakout detection owing to its sensitivity [37]. Many studies have adopted this test, including those analyzing precipitation and temperature [38,39], as it can detect a mutation points within specific periods of time in many climate records [40].

The regions affected by TCs change over time and are subject to various TC intensities and tracks. Therefore, it is necessary to consider the long-term "ramping-up point" in each river basin and to comprehensively determine propagation of the TC-induced rainfall. However, the Pettitt test can be applied to discern a significant transformation in the mean of a time series without prior knowledge of the ramp-up point. The nonparametric statistic K_T, the ramping-up within period T, is defined as follows (Equation (3)):

$$K_T = \text{Max}|U_{t,T}| \tag{3}$$

$$U_{t,T} = \sum_{i=1}^{t} \sum_{j=t+1}^{T} \text{sign}(x_i - x_j), \ 1 \le t < T \tag{4}$$

where $U_{t,\,T}$ is a statistical index, and x is a random variable.

Subsequently, for the significance analysis, the correlative confidence level ρ and the roughly critical value p were defined as follows (Equation (5)):

$$\rho = 2 \exp\left(\frac{-6\, K_T^2}{T^2 + T^3} \right) \tag{5}$$

$$p = 1 - \rho \tag{6}$$

3. Analysis Results

3.1. Characteristics of TCs under Different Types of El Niños

3.1.1. Magnitude of TCs

To analyze the characteristics of TCs, we discussed variations in TCs during the developing and decaying years of MCP, MEP, and SBW El Niño events and compared them with long-term averages to help predict the behavior of TCs occurring under different ENSO types and phases.

Figure 2 shows the distribution of TC data observed across the WNP and its probability density during the developing and decaying years of ENSO. TCs were recorded, with 186, 140, and 86 events, respectively, during MCP, MEP, and SBW decaying years. TCs occurred more frequently during MEP events than during the other El Niño types with an average of 28 times a year. During SBW decaying years, the number of annual TCs was above average. To explore the varying effects throughout El Niño periods, we studied both the emerging and decaying years. During MCP, MEP, and SBW developing years, 217, 130, and 112 TC events were recorded, respectively. In addition, there was little interannual variation within the El Niño types, and the number of TC events in MCP and SBW decaying years was greater than that during developing years. However, the number of TCs increased significantly during MEP developing years.

Because wind speed data were only available for 1977, wind speed statistics for MEP events were limited. In 1977 (characterized by MEP decay), only 21 TCs were produced, and no data were available for emerging years. There were 81 TC events during the three SBW events in 1977. The wind speed data indicated that these TCs were weaker than the long-term average and that anomalies were more concentrated in the range of -30 to 20 knots in the decaying years of MEP and SBW events. However, TCs occurring during SBW emerging years had significantly faster wind speeds. TCs during MCP events were little impacted by El Niño phases and featured slightly higher wind speeds than the long-term average. TCs with relatively low wind speeds were observed during SBW decaying years. Central pressure anomaly changes were relatively minor during MCP events, and pressure remained slightly below the multiyear average. TC intensity was obviously greater during SBW emerging years. In MEP and SBW decaying years, most TCs were weaker than average, though a few extremely strong TCs occurred. In terms of duration, the lifespan of TCs ranged widely during El Niño emerging years. The differences in duration were relatively significant in SBW years but were less marked during MCP events.

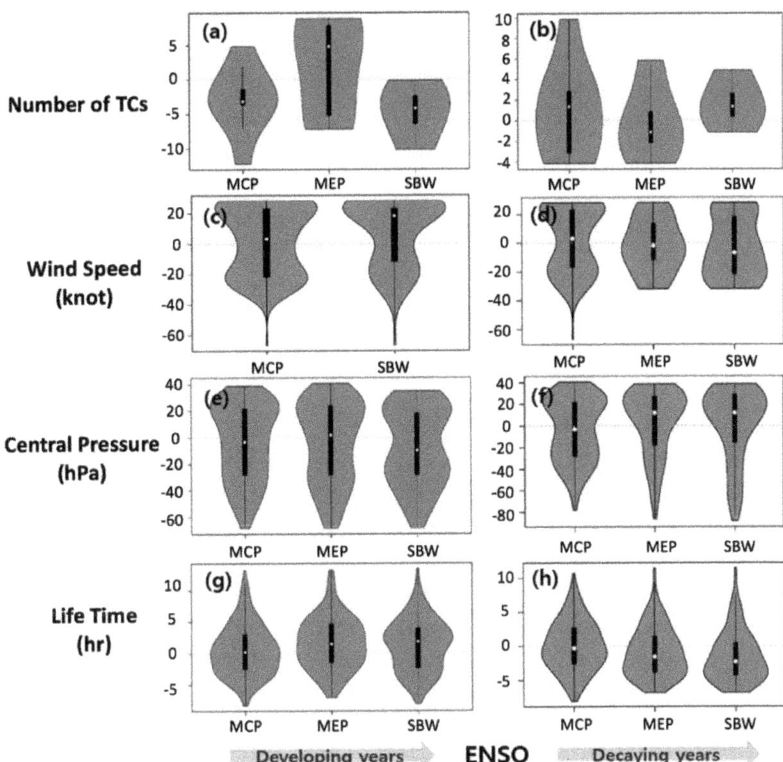

Figure 2. Violin plots of tropical cyclone (TC) activities and their probability during different types of El Niño–Southern Oscillation (ENSO) events. (**a,b**) show the annual number of TCs, (**c,d**) indicate wind speed anomaly patterns of TCs, (**e,f**) show the central pressure anomaly, and (**g,h**) indicate lifetime anomalies during developing and decaying years. For each panel, the violin plot shows the kernel probability density of the data, including a dot to mark the median and a thick line to indicate the interquartile range of each dataset. MCP: moderate central Pacific, MEP: moderate eastern Pacific, SBW: strong basin-wide.

3.1.2. Genesis Positions and Tracks of TCs

Figure 3 shows the genesis positions of TCs in the WNP Ocean region during the different types of El Niño years. The genesis positions in MCP emerging (Figure 3a, average TC genesis: 144.2° E, 13.2° N) and decaying years (Figure 3b, average TC genesis: 143.1° E, 13.7° N) were similar to those in average years (average TC genesis: 141.8° E, 13.7° N). In MEP developing years, genesis locations were relatively concentrated at lower latitudes (Figure 3c, average TC genesis: 145.4° E, 11.7° N). However, during MEP decay, the genesis locations (Figure 3d, average TC genesis: 142.5° E, 13.2° N) were close to their long-term normal distribution. In contrast, the genesis positions in SBW decaying years tended to be more westerly and closer to China (Figure 3f, average TC genesis: 134.8° E, 13.2° N).

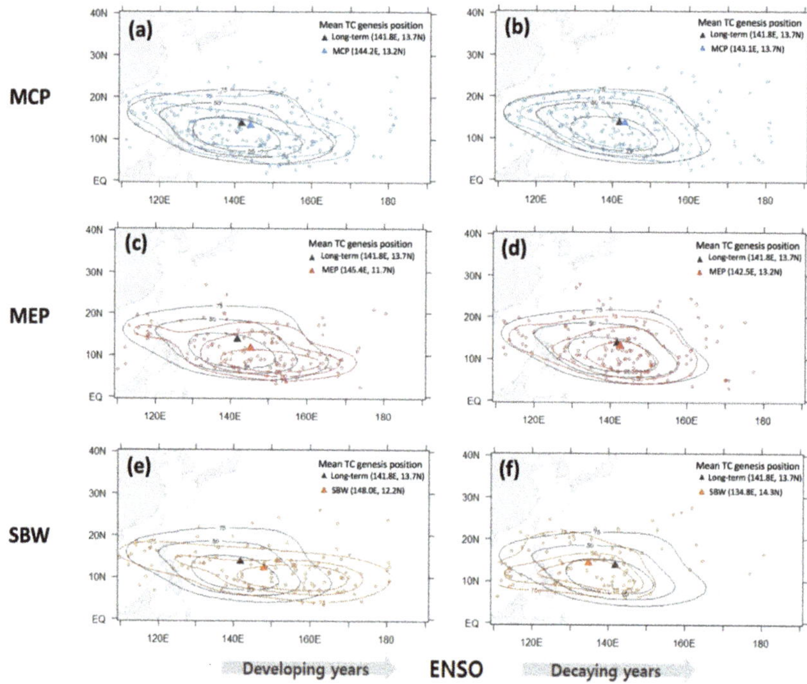

Figure 3. Genesis positions of tropical cyclones (TCs) during the development and decay of moderate central Pacific (MCP), moderate eastern Pacific (MEP), and strong basin-wide (SBW) El Niño events. The black contour lines show the long-term trends in TC genesis locations (1961–2017). The TC genesis position is defined as the first location at which a TC was recorded in historical data provided by the Regional Specialized Meteorological Center in Tokyo, Japan. (**a,b**) moderate central Pacific (MCP); (**c,d**) moderate eastern Pacific (MEP); (**e,f**) strong basin-wide (SBW).

Figure 4 shows the density circle of recurving locations during the decay period of three types of El Niño events. Overall, 52, 30, and 14 TCs displayed recurving features (during MCP, MEP, and SBW events, respectively). During SBW events, these points are located close to China, and the contours were wider during MCP events. Recurving points during MEP decay were located in lower latitude zones.

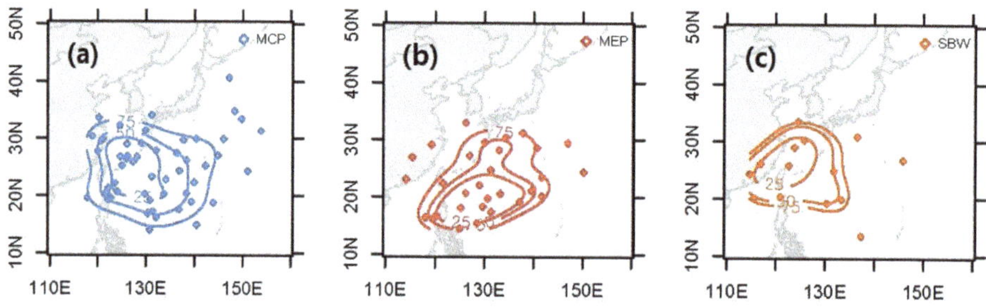

Figure 4. Map of the density circle of tropical cyclone recurving positions during the decay periods of (**a**) moderate central Pacific (MCP), (**b**) moderate eastern Pacific (MEP), and (**c**) strong basin-wide (SBW) El Niño events.

Figure 5 shows the subtropical high, represented by 5880 gpm, in the WNP region for various El Niño conditions. A close relationship exists between the subtropical high and TCs as streams expand and convection occurs [41]. TC expansion and recurving locations were affected by the pressure zone. During MEP decaying years, the subtropical high was distinguished by a limited pressure range. As extreme events propagated, the range of the subtropical high widened during MCP and SBW events. Regarding track density, TCs generated in the WNP usually originated near the southern edge of the subtropical high and moved along its periphery. During MCP and SBW decaying years, the subtropical high was slightly stronger than its long-term average.

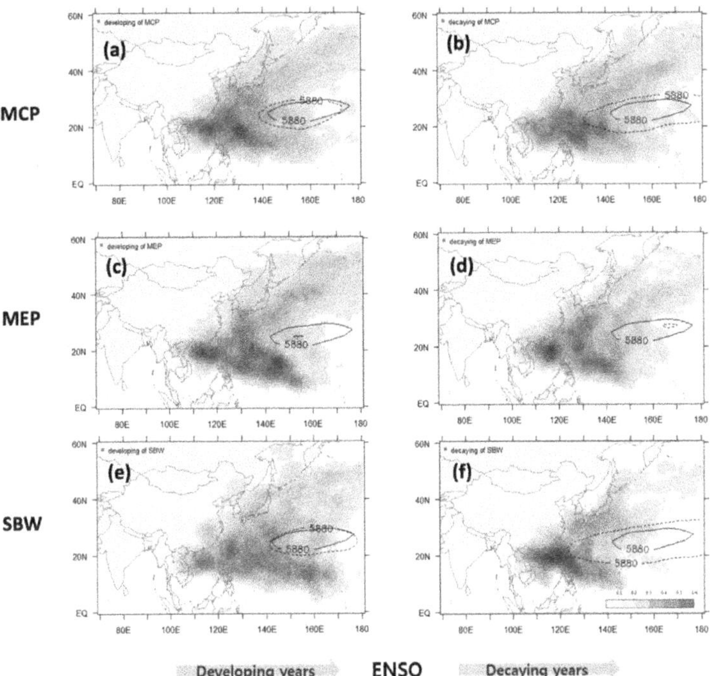

Figure 5. Comparison of subtropical highs and tropical cyclone behavior under different types of El Niño–Southern Oscillation (ENSO) events during the developing and decaying years, ((**a**,**b**) moderate central Pacific (MCP); (**c**,**d**) moderate eastern Pacific (MEP); (**e**,**f**) strong basin-wide (SBW)). The solid line shown in each figure indicates the long-term average position of the western North Pacific (WNP) subtropical high (represented by 5880 gpm). Dashed lines indicate different types of ENSO events. The grey shading marks the average path of the tropical cyclones for each case.

3.2. TC-Induced Rainfall over China
Intensity of TC-Induced Rainfall

Figure 6 shows the proportion of TC-induced rainfall (PTCR) in each river basin throughout China. The Pearl, Yangtze, and Southeastern river basins were the regions most affected by TCs. During MCP events, the PTCR was less than 11.6% (long-term average: 35.92% ≥ MCP years: 31.74%) in the Pearl river basin compared to the long-term normal. In contrast, the PTCR was shown to be large by 26.9% (long-term average: 14.96% ≥ MCP years: 18.99%) and 6.2% (long-term average: 19.51% ≥ MCP years: 20.72%), respectively, in the Southeast and Yangtze river basins. During MEP events, the PTCR in these three river basins dropped slightly compared to the average year. During SBW events, however, the PTCR was larger than usual in all these three basins, indicating a clear increase in rainfall caused by typhoons.

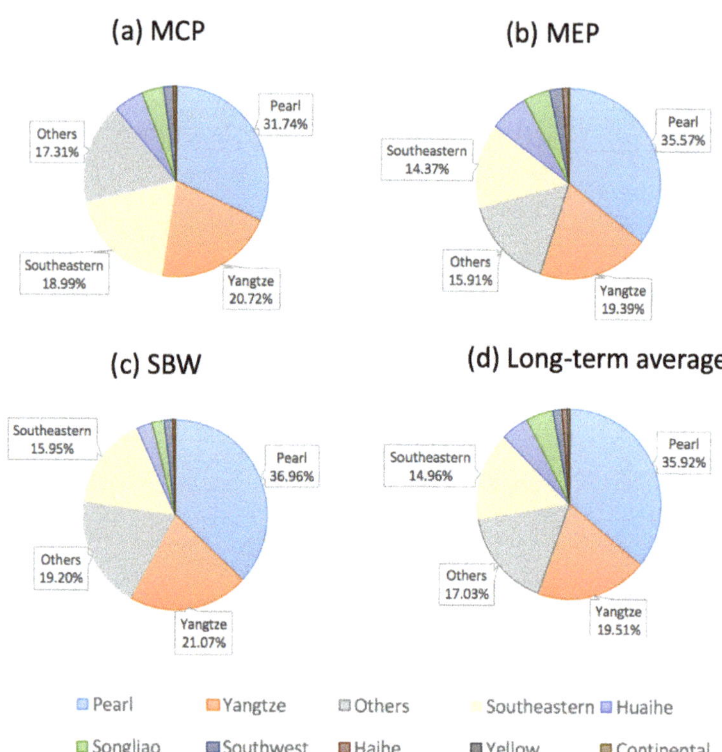

Figure 6. Proportion of tropical cyclone-induced precipitation amount in each river basin of China. (**a–c**) show results for moderate central Pacific (MCP), (**b**) moderate eastern Pacific (MEP), and (**c**) strong basin-wide (SBW) El Niño events, respectively, and (**d**) shows the long-term average.

Figure 7 compares TC-induced composite PRCPTOT anomalies with the long-term values observed at each station throughout China. The four significance levels were determined based on Welch's *t*-test. Total seasonal precipitation was concentrated in the Yangtze, Pearl, Hong Kong, and Southeastern river basins during MCP development. Then, rainfall moved to the coastal areas of Huaihe and Southeast. During MEP El Niño events, rainfall continued to fall in the Southwest, Pearl, and Songliao river basins. In addition, during SBW events when rain moved to the Yangtze and Hong Kong river basins, more rain was concentrated in the Huaihe, Haihe, and Songliao regions. During the emergence of MEP and SBW events and the decay of MEP events, the Songliao River Basin experienced more TC-induced rainfall than usual. TC-induced rainfall was slightly reduced during SBW events, and TC-induced rainfall increased in the Southeast during MCP decay. The maximum rainfall over China noticeably reduced. However, there was slightly more rain in Hong Kong, and Songliao experienced slightly more rain under MEP decay. The Yangtze and Pearl regions received relatively more rain in almost all cases. Therefore, the results suggest that TC-induced rainfall varied with TC tracks.

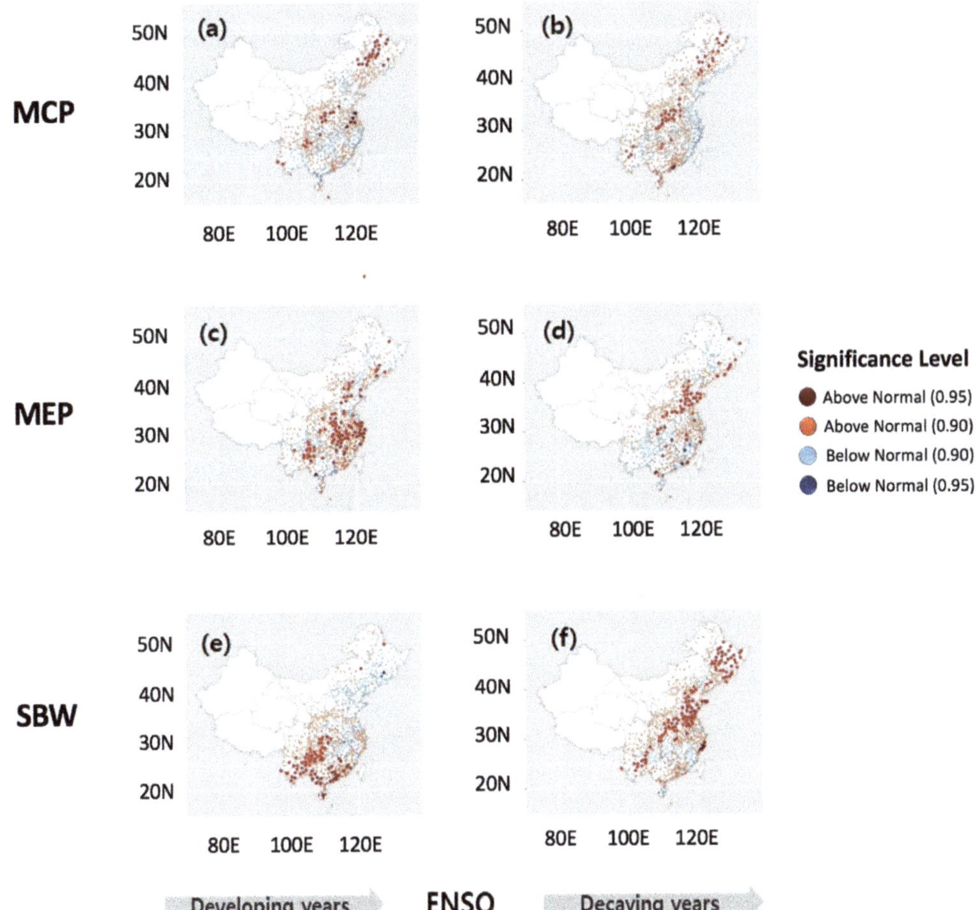

Figure 7. TC-induced composite PRCPTOT anomalies compared to the long-term averages. The red dots indicate above-normal conditions, whereas the blue dots indicate below-normal conditions. In addition, the light (dark) color groups indicate 90% (95%) significance levels for pattern changes. MCP: moderate central Pacific, MEP: moderate eastern Pacific, SBW: strong basin-wide. (**a,b**) moderate central Pacific (MCP); (**c,d**) moderate eastern Pacific (MEP); (**e,f**) strong basin-wide (SBW).

Figure 8 shows the maximum number of consecutive TC-induced precipitation days (CTPD), which is presented as a composite anomaly compared to the long-term average. Each CTPD represents a day consistently affected by TCs. During MCP development, CTPD distribution results were not clear, whereas CTPD expanded to the inland and northeastern China during MCP decaying years. However, SBW development was associated with distinct CTPDs; the southwestern region was affected by TCs over a long period, with the eastern and northeastern regions being affected significantly fewer days. During El Niño decay, coastal areas were relatively more affected by TCs during MCP events, and the Pearl River Basin was more affected during SBW events. CTPD was distributed through central and southern China with a clear boundary in the Songliao River Basin.

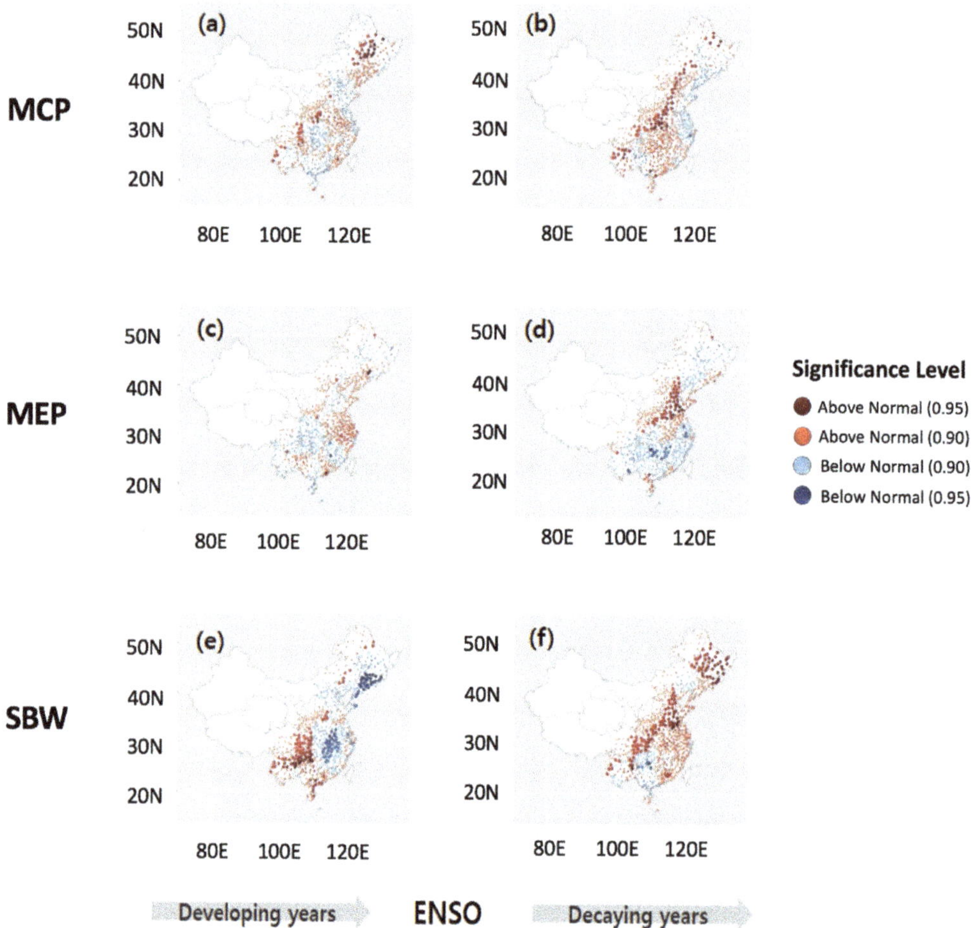

Figure 8. Composite consecutive tropical cyclone-induced precipitation days (CTPD) anomalies compared to long-term averages. The red dots indicate above normal results, whereas the blue dots indicate below normal results. The light-(dark-)colored shaded indicates change at a 90% (95%) significance level. MCP: moderate central Pacific, MEP: moderate eastern Pacific, SBW: strong basin-wide. (**a,b**) moderate central Pacific (MCP); (**c,d**) moderate eastern Pacific (MEP); (**e,f**) strong basin-wide (SBW).

Figure 9 shows composite anomalies in the average number of days per year affected by TCs. Under extreme El Niño events, slightly fewer days were affected by TCs in China. For inland areas, anomalies generally comprised 10 or fewer days and never exceeded 12. A distinct pattern was observed over the Songliao River Basin, indicating the impact of TCs could extend into northeast China. During SBW years, the range of areas impacted by TCs was narrow, especially in the developing years. However, the Yangtze River Basin was affected by TCs for approximately six days.

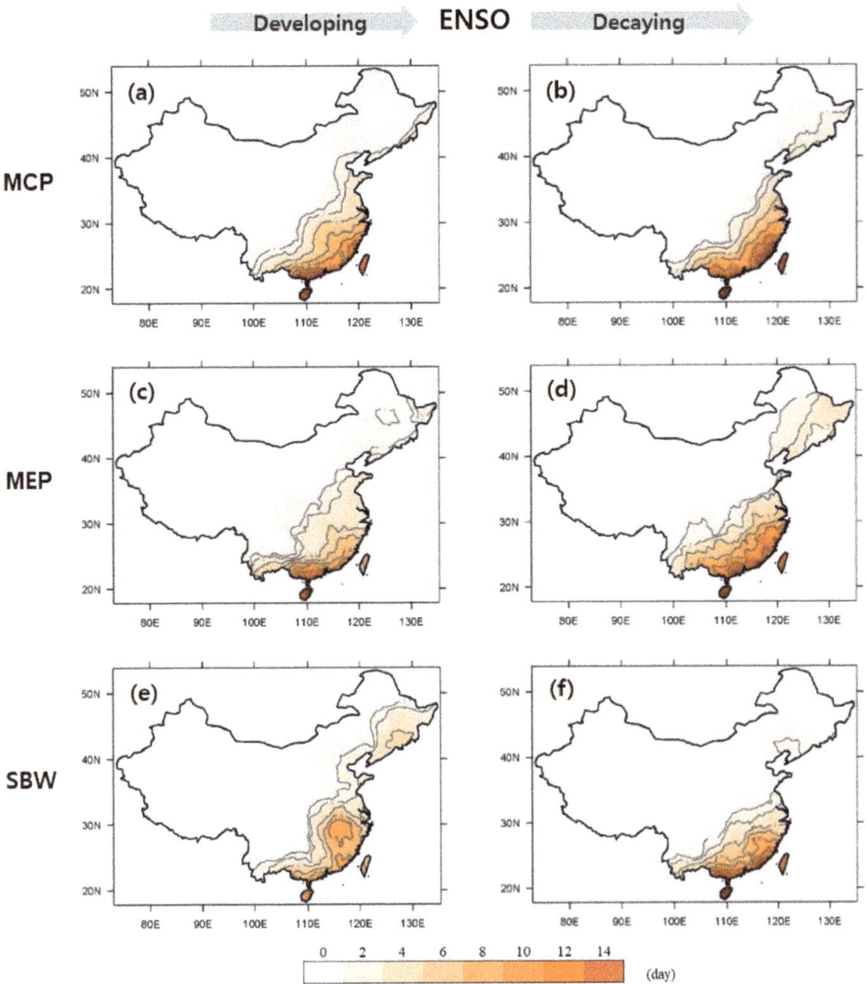

Figure 9. Composite anomaly map showing the average number of days per year affected by tropical cyclones. The contour lines, as they extend from west to east, indicate areas affected on 1, 2, 4, 6, 8, 10, and 12 days per year. Dark orange areas denote areas that were affected by anomalies in all years, which mainly include the Pearl and Southeastern river basins and offshore islands. MCP: moderate central Pacific, MEP: moderate eastern Pacific, SBW: strong basin-wide. (**a,b**) moderate central Pacific (MCP); (**c,d**) moderate eastern Pacific (MEP); (**e,f**) strong basin-wide (SBW).

4. Summary and Conclusions

El Niño events significantly influence TC characteristics and behaviors in the WNP, as they cause changes in ocean and atmospheric circulation. This statistical study investigated how ENSO types (MCP, MEP, and SBW) affected various TC properties and rainfall over different regions in China to support the development of more accurate prediction models.

This study first analyzed the characteristics of TCs and classified them into TC strength, frequency, and tracks. In terms of TC strength, the MCP El Niño type was associated with a slightly larger than average SSTA in both emerging to decaying phases. However, the strength of TCs declined in the MEP and SBW processes. Among them, the SBW showed more variability. For TC frequency, the annual numbers of TCs increased significantly to reach their maximum, whereas other TCs had opposite patterns. During MEP and SBW development, compared to the long-term distribution, TC track positions were distributed eastward, especially in the case of the SBW. In the decaying years, TC events were similar to long-term distributions during MCP and MEP events, and TCs tended to occur close to the WNP region in SBW. For recurving positions, the situation was similar in that the range of genesis points approached China during SBW years.

TC-induced rainfall anomalies varied by region and ranged from relatively heavy to weak. In all cases, the Pearl River Basin and Southern river basins experienced a high proportion of TC-induced rainfall, especially in coastal areas. Rainfall induced under the three El Niño types is compared to the average TC-induced rainfall, which shows extreme rainfall occurred more often during El Niño events. During MCP events, the proportion of TC-induced rainfall was similar to the average distribution in all years. During MEP events, rainfall was more extensive, spreading to the Songliao River Basin and contributing to TC-induced rainfall in some areas of the Yangtze River Basin.

With respect to TC-induced precipitation indices, two indicators illustrated PRCPTOT, CTPD, and other characteristics in six situations, thereby representing the three types of developing and decaying years of El Niño. Statistical results showed that there was TC-induced heavy rainfall in southern China and coastal areas of the Huaihe River Basin during the developing years of MCP. TC-induced heavy rain occurred in eastern and northeastern China in SBW, rain during MEP years was relatively weak, and the precipitation in the Songliao region was relatively strong. In the decaying years, heavy rain was distributed in the eastern coastal cities, the Southwest and Songliao river basins, and the Yangtze River Basin in MCP, MEP, and SBW, respectively. The frequency showed a similar pattern as TC-induced PRCPTOT, except that the consistent impacts of TCs covered large areas in southern China during the years of MEP. With respect to geographical frequency, TC-induced rainfall at the stage of the SBW El Niño events was distinguished primarily by the narrow range of areas affected by TCs in the Pearl and Southeastern regions of China.

For SBW events, warming started in the western Pacific and initially spread eastward. They have the pronounced westerly anomalies happening during the winter and spring over the western Pacific, possibly reflecting frequent westerly wind-burst events. During the decaying years of SBW events, more TCs occurred in the eastern Philippines and the South China Sea than usual, and the genesis positions of TCs tended to be more westerly and closer to China (Figure 3f, average TC genesis: 134.8° E, 13.2° N), and they moved towards the Pearl River and Southeastern river basins. One of the possible physical processes is that the difference in SSTs between the eastern Pacific and the central Pacific causes westerly anomaly winds, further extending the WNP monsoon trough to the east and resulting in relatively increased TC events in the east WNP. In particular, during SBW events, a subtropical high maintains stronger than usual, and convection activities surrounding the subtropical western pacific stimulate a Rossby wave that propagates from the South China Sea to midlatitudes, causing an anomalous easterly steering flow in the Pearl River and Southeastern river basins. Although numerical studies are needed to confirm the reliability of the results we attained, which were limited by our relatively small sample sizes, this diagnostic study presented a comparative analysis of the characteristics of WNP TCs occurring under three types of El Niño events (MCP, MEP, and SBW) and their local impacts on precipitation.

Though changes in El Niño intensity under anthropogenic warming are critical considerations, current prediction models remain inadequate. For example, they do not distinguish between strong and moderate El Niño events, making it difficult to predict future changes in the intensity of El Niño. We anticipate that our findings will help improve prediction models by promoting a better understanding of the relationship between El Niño and TC characteristics. In addition, the results of this study have helped determine typical rainfall patterns at different stages of various El Niño events and thus could support more accurate rainfall predictions under future climatic conditions in China.

Author Contributions: Conceptualization, Y.L. and J.-S.K.; formal analysis, Y.L.; methodology, Y.L. and J.-S.K.; resources, J.-S.K. and S.-K.Y.; writing—original draft preparation, Y.L., J.-S.K., and J.-H.L.; writing—review and editing, J.-S.K., S.-K.Y., and L.X. All authors have read and agreed to the published version of the manuscript.

Funding: This research is supported by the National Natural Science Foundation of China (NSFC Grant Nos. 41890822 and 51525902). The second author, Dr. Yoon, was partially supported by the Seoul Institute of Technology (2020-AB-003). In addition, this work was supported by Korea Environment Industry & Technology Institute (KEITI) though Water Management Research Program, funded by Korea Ministry of Environment (MOE) (79616).

Acknowledgments: We appreciate the support of the State Key Laboratory of Water Resources and Hydropower Engineering Science, Wuhan University. The authors thank the Editor and three anonymous reviewers for their insightful comments and suggestions.

Conflicts of Interest: The authors declare no conflict of interest.

References

1. Lin, Y.; Zhao, M.; Zhang, M. Tropical cyclone rainfall area controlled by relative sea surface temperature. *Nat. Commun.* **2015**, *6*, 1–7. [CrossRef]
2. Lau, K.-M.; Zhou, Y.P.; Wu, H.-T. Have tropical cyclones been feeding more extreme rainfall? *J. Geophys. Res. Atmos.* **2008**, *113*. [CrossRef]
3. Zhao, L.; Bai, X.; Qi, D.; Xing, C. BMA probability quantitative precipitation forecasting of land-falling typhoons in south-east China. *Front. Earth Sci.* **2019**, *13*, 758–777. [CrossRef]
4. Zhao, H.; Wang, C. On the relationship between ENSO and tropical cyclones in the western North Pacific during the boreal summer. *Clim. Dyn.* **2019**, *52*, 275–288. [CrossRef]
5. Deser, C.; Alexander, M.A.; Xie, S.-P.; Phillips, A.S. Sea surface temperature variability: Patterns and mechanisms. *Annu. Rev. Mar. Sci.* **2010**, *2*, 115–143. [CrossRef] [PubMed]
6. Shuqiu, L.X.Y. El Niño and rainfall during the flood season (June-August) in China. *Acta Meteorol. Sin.* **1993**, *51*, 434–441.
7. Zhang, X.; Alexander, L.; Hegerl, G.C.; Jones, P.; Tank, A.K.; Peterson, T.C.; Trewin, B.; Zwiers, F.W. Indices for monitoring changes in extremes based on daily temperature and precipitation data. *Wiley Interdiscip. Rev. Clim. Chang.* **2011**, *2*, 851–870. [CrossRef]
8. Zhang, Q.; Lai, Y.; Gu, X.; Shi, P.; Singh, V.P. Tropical cyclonic rainfall in China: Changing properties, seasonality, and causes. *J. Geophys. Res. Atmos.* **2018**, *123*, 4476–4489. [CrossRef]
9. Yonekura, E.; Hall, T.M. A statistical model of tropical cyclone tracks in the western North Pacific with ENSO-dependent cyclogenesis. *J. Appl. Meteorol. Clim.* **2011**, *50*, 1725–1739. [CrossRef]
10. Hirata, H.; Kawamura, R. Scale interaction between typhoons and the North Pacific subtropical high and associated remote effects during the Baiu/Meiyu season. *J. Geophys. Res. Atmos.* **2014**, *119*, 5157–5170. [CrossRef]
11. Sun, Y.; Zhong, Z.; Yi, L.; Li, T.; Chen, M.; Wan, H.; Wang, Y.; Zhong, K. Dependence of the relationship between the tropical cyclone track and western Pacific subtropical high intensity on initial storm size: A numerical investigation. *J. Geophys. Res. Atmos.* **2015**, *120*, 11–451. [CrossRef]
12. Chen, X.; Zhong, Z.; Lu, W. Association of the poleward shift of East Asian subtropical upper-level jet with frequent tropical cyclone activities over the western North Pacific in summer. *J. Clim.* **2017**, *30*, 5597–5603. [CrossRef]
13. Cai, M.; Ding, Y.; Jiang, Z. Extreme Precipitation Experimentation over Eastern China Based on L-moment Estimation. *Plateau Meteorol.* **2007**, *26*, 012.
14. Corporal-Lodangco, I.L.; Leslie, L.M.; Lamb, P.J. Impacts of ENSO on Philippine tropical cyclone activity. *J. Clim.* **2016**, *29*, 1877–1897. [CrossRef]
15. Camargo, S.J.; Sobel, A.H. Western North Pacific tropical cyclone intensity and ENSO. *J. Clim.* **2005**, *18*, 2996–3006. [CrossRef]
16. Colbert, A.J.; Soden, B.J.; Kirtman, B.P. The impact of natural and anthropogenic climate change on western North Pacific tropical cyclone tracks. *J. Clim.* **2015**, *28*, 1806–1823. [CrossRef]
17. Liu, Z.; Chen, X.; Sun, C.; Cao, M.; Wu, X.; Lu, S. Influence of ENSO Events on Tropical Cyclone Activity over the Western North Pacific. *J. Ocean Univ. China* **2019**, *18*, 784–794. [CrossRef]
18. Okumura, Y.M. ENSO diversity from an atmospheric perspective. *Curr. Clim. Chang. Rep.* **2019**, *5*, 245–257. [CrossRef]
19. Cai, W.; McPhaden, M.J.; Grimm, A.M.; Rodrigues, R.R.; Taschetto, A.S.; Garreaud, R.D.; Dewitte, B.; Poveda, G.; Ham, Y.-G.; Santoso, A. Climate impacts of the El Niño–Southern Oscillation on South America. *Nat. Rev. Earth Environ.* **2020**, *1*, 215–231. [CrossRef]
20. Wang, B.; Luo, X.; Yang, Y.-M.; Sun, W.; Cane, M.A.; Cai, W.; Yeh, S.-W.; Liu, J. Historical change of El Niño properties sheds light on future changes of extreme El Niño. *Proc. Natl. Acad. Sci. USA* **2019**, *116*, 22512–22517. [CrossRef]
21. Li, Y.; Ma, B.; Feng, J.; Lu, Y. Influence of the strongest central Pacific El Niño–Southern Oscillation events on the precipitation in eastern China. *Int. J. Clim.* **2019**, *39*, 3076–3090. [CrossRef]

22. Lu, A.; Jia, S.; Yan, H.; Wang, S. El Nino-Southern Oscillation and water resources in headwaters region of the Yellow River: Links and potential for forecasting. *Hydrol. Earth Syst. Sci. Discuss.* **2010**, *7*, 8521–8543. [CrossRef]

23. Kim, J.-S.; Jain, S.; Yoon, S.-K. Warm season streamflow variability in the Korean Han River Basin: Links with atmospheric teleconnections. *Int. J. Clim.* **2012**, *32*, 635–640. [CrossRef]

24. Kim, J.-S.; Jain, S. Precipitation trends over the Korean peninsula: Typhoon-induced changes and a typology for characterizing climate-related risk. *Environ. Res. Lett.* **2011**, *6*, 034033. [CrossRef]

25. Qian, W.; Lin, X. Regional trends in recent precipitation indices in China. *Meteorol. Atmos. Phys.* **2005**, *90*, 193–207. [CrossRef]

26. Laing, A.; Evans, J.L. *Introduction to Tropical Meteorology*; Educational Material from the COMET Program; COMET MetEd: Boulder, CO, USA, 2011.

27. Kalnay, E.; Kanamitsu, M.; Kistler, R.; Collins, W.; Deaven, D.; Gandin, L.; Iredell, M.; Saha, S.; White, G.; Woollen, J.; et al. The NCEP/NCAR 40-year reanalysis project. *Bull. Amer. Meteor. Soc.* **1996**, *77*, 437–470. [CrossRef]

28. Matheron, G. Principles of geostatistics. *Econ. Geol.* **1963**, *58*, 1246–1266. [CrossRef]

29. Le, N.D.; Zidek, J.V. *Statistical Analysis of Environmental Space-Time Processes*; Springer Science & Business Media: Berlin, Germany, 2006.

30. Cressie, N. The origins of kriging. *Math. Geol.* **1990**, *22*, 239–252. [CrossRef]

31. Delhomme, J.P. Kriging in the hydrosciences. *Adv. Water Resour.* **1978**, *1*, 251–266. [CrossRef]

32. Su, S.; Lin, A.; Liu, Q. The application of ordinary Kriging method in spatial interpolation. *J. Jiangnan Univ. (Nat. Sci. Ed.)* **2004**, *3*, 18–21.

33. Welch, B.L. The generalization of 'Student's' problem when several different population variances are involved. *Biometrika* **1947**, *34*, 28–35. [CrossRef] [PubMed]

34. Yin, X. *The Principle and Operation of Econometrics*; ChongQing University Press: ChongQing, China, 2009; Volume 8.

35. Machiwal, D.; Jha, M.K. *Hydrologic Time Series Analysis: Theory and Practice*; Springer Science & Business Media: Berlin, Germany, 2012.

36. Pettitt, A.N. A non-parametric approach to the change-point problem. *J. R. Stat. Soc. Ser. C (Appl. Stat.)* **1979**, *28*, 126–135. [CrossRef]

37. Wijngaard, J.B.; Klein Tank, A.M.G.; Können, G.P. Homogeneity of 20th century European daily temperature and precipitation series. *Int. J. Clim. J. R. Meteorol. Soc.* **2003**, *23*, 679–692. [CrossRef]

38. Smadi, M.M.; Zghoul, A. A sudden change in rainfall characteristics in Amman, Jordan during the mid 1950s. *Am. J. Environ. Sci.* **2006**, *2*, 84–91. [CrossRef]

39. Dhorde, A.G.; Zarenistanak, M. Three-way approach to test data homogeneity: An analysis of temperature and precipitation series over southwestern Islamic Republic of Iran. *J. Indian Geophys. Union* **2013**, *17*, 233–242.

40. Jaiswal, R.K.; Lohani, A.K.; Tiwari, H.L. Statistical analysis for change detection and trend assessment in climatological parameters. *Environ. Process.* **2015**, *2*, 729–749. [CrossRef]

41. Kim, J.-S.; Kim, S.T.; Wang, L.; Wang, X.; Moon, Y.-I. Tropical cyclone activity in the northwestern Pacific associated with decaying Central Pacific El Ninos. *Stoch. Environ. Res. Risk Assess.* **2016**, *30*, 1335–1345. [CrossRef]

 *atmosphere*

Article

Complexity of Forces Driving Trend of Reference Evapotranspiration and Signals of Climate Change

Mohammad Valipour [1,2,*], Sayed M. Bateni [1], Mohammad Ali Gholami Sefidkouhi [3], Mahmoud Raeini-Sarjaz [3] and Vijay P. Singh [4]

[1] Department of Civil and Environmental Engineering and Water Resources Research Center, University of Hawaii at Manoa, Honolulu, HI 96822, USA; smbateni@hawaii.edu

[2] Center of Excellence for Climate Change Research/Department of Meteorology, King Abdulaziz University, Jeddah 21589, Saudi Arabia

[3] Department of Water Engineering, Sari Agricultural Sciences and Natural Resources University, Sari 382220, Iran; magholamis@yahoo.com (M.A.G.S.); raeini@yahoo.com (M.R.-S.)

[4] Department of Biological and Agricultural Engineering and Zachry Department of Civil Engineering, Texas A and M University, 321 Scoates Hall, 2117 TAMU, College Station, TX 77843-2117, USA; vsingh@tamu.edu

[*] Correspondence: valipour@hawaii.edu

Received: 1 September 2020; Accepted: 5 October 2020; Published: 10 October 2020

Abstract: Understanding the trends of reference evapotranspiration (ET_0) and its influential meteorological variables due to climate change is required for studying the hydrological cycle, vegetation restoration, and regional agricultural production. Although several studies have evaluated these trends, they suffer from a number of drawbacks: (1) they used data series of less than 50 years; (2) they evaluated the individual impact of a few climatic variables on ET_0, and thus could not represent the interactive effects of all forces driving trends of ET_0; (3) they mostly studied trends of ET_0 and meteorological variables in similar climate regions; (4) they often did not eliminate the impact of serial correlations on the trends of ET_0 and meteorological variables; and finally (5) they did not study the extremum values of meteorological variables and ET_0. This study overcame the abovementioned shortcomings by (1) analyzing the 50-year (1961–2010) annual trends of ET_0 and 12 meteorological variables from 18 study sites in contrasting climate types in Iran, (2) removing the effect of serial correlations on the trends analysis via the trend-free pre-whitening approach, (3) determining the most important meteorological variables that control the variations of ET_0, and (4) evaluating the coincidence of annual extremum values of meteorological variables and ET_0. The results showed that ET_0 and several meteorological variables (namely wind speed, vapor pressure deficit, cloudy days, minimum relative humidity, and mean, maximum and minimum air temperature) had significant trends at the confidence level of 95% in more than 50% of the study sites. These significant trends were indicative of climate change in many regions of Iran. It was also found that the wind speed (WS) had the most significant influence on the trend of ET_0 in most of the study sites, especially in the years with extremum values of ET_0. In 83.3% of the study sites (i.e., all arid, Mediterranean and humid regions and 66.7% of semiarid regions), both ET_0 and WS reached their extremum values in the same year. The significant changes in ET_0 due to WS and other meteorological variables have made it necessary to optimize cropping patterns in Iran.

Keywords: reference evapotranspiration; climate change; drought; meteorological extremes; climatic variables; wind speed

1. Introduction

Assessment of changes in reference evapotranspiration (ET_0) is required in climate change studies, agricultural and forest meteorology, irrigation scheduling, surface water balance, drought analysis, long-term decision-making in food and water security policies, and optimum allocation of water resources [1–7]. Evaluating the trends of meteorological variables may help determine the effects of major factors on ET_0 and climate change [8,9].

Dadaser-Celik et al. [10] evaluated the 32-year trend of ET_0 in Turkey. Analysis of climatic data showed an upward trend in air temperature, and downward trends in wind speed and relative humidity in Turkey. Changes in these three variables explained the majority of variations in ET_0. Song et al. [11] assessed the 46-year trend of ET_0 in the North China Plain. Their results indicated that the downward trends of net radiation and wind speed had a bigger impact on ET_0 compared to the upward trends of maximum and minimum air temperature. Wang et al. [12] characterized the 31-year trend of ET_0 in the western Heihe River Basin in China. They found that wind speed and sunshine duration were the two key meteorological variables that decreased ET_0.

Darshana et al. [13] investigated the 30-year trend of ET_0 in the Tons River Basin in central India. Their outcomes showed that maximum air temperature and net radiation had a stronger impact on ET_0 compared to minimum air temperature and relative humidity. Zongxing et al. [14] studied the 49-year trend of ET_0 in the south-west of China. The decrease in wind speed was the main driving force for the reduction of ET_0. This happened because the lower wind speed raised the vapor pressure, which ultimately reduced the evaporative demand of atmosphere. Li et al. [15] examined the 46-year trend of ET_0 in the Upper Mekong River Basin. They showed that sunshine duration had a more significant (at the confidence level 95%) effect on ET_0 compared to air temperature, relative humidity, and wind speed. Gao et al. [16] evaluated the 45-year trend of ET_0 in China, and found that sunshine duration, wind speed, and relative humidity had a more important influence on ET_0 than air temperature. Zhang et al. [17] assessed the changes in ET_0 and its controlling factors in China. They indicated that maximum air temperature, relative humidity, and wind speed affected ET_0. A number of studies (e.g., [8,9,18–20]) showed that the increasing and decreasing trends of ET_0 were mainly due to the increase in air temperature and decrease in wind speed, respectively.

Tabari et al. [21,22] evaluated the 40-year (1966–2005) trend of ET_0 in the west and southwest of Iran with arid and semiarid climates. They studied the effect of air temperature, relative humidity, vapor pressure, wind speed, and rainfall on ET_0, and found that wind speed had the most influence on ET_0. Unfortunately, they did not consider the impact of serial correlations (autocorrelation) on the trends of ET_0 and meteorological variables. Kousari and Ahani [23] assessed trends of ET_0 and six meteorological variables (mean, minimum and maximum air temperature, relative humidity, wind speed and sunshine hours) in three climate regions (arid, semiarid and humid) of Iran during 1975–2005. They did not take into account the influence of serial correlations on the trends. Shadmani et al. [24] evaluated the 41-year (1965–2005) trend of ET_0 from 11 weather stations in arid regions of Iran but did not specify which meteorological variables controlled trend of ET_0.

The existing studies (1) evaluated the impact of various climatic variables on ET_0 and (2) showed the signals of climate change in Iran based on upward/downward trends of ET_0 and meteorological variables [8,9,20–23]. However, they suffer from the following shortcomings: (1) they used data series of less than 50 years which are suitable only for evaluating climate variability and not climatic change. At least a 50-year period is necessary to study climate change [25–32]. Borman [26] used a 50-year period to analyze the sensitivity of ET_0 to climate change in Germany. Kingston et al. [27] found a considerable uncertainty in evaluating the changes of ET_0 due to climate change using 30-year data. Several studies assessed the response of ET_0 to climate change in China using 50-year data [28–31]. Hence, there is a consensus in the literature to use at least a 50-year period to evaluate the trend of ET_0 due to climate change. (2) They evaluated the individual impact of a few climatic variables on ET_0, and thus could not represent the interactive effects of all forces driving trends of ET_0. (3) They mostly studied trends of ET_0 and meteorological variables in similar climate regions. (4) They often did not eliminate the impact of serial correlations (autocorrelation) on the trends of ET_0 and meteorological variables. Finally (5) they did not assess if the extremum values of meteorological variables and ET_0 occur simultaneously. This study overcame the abovementioned drawbacks by analyzing the 50-year (1961–2010) trends of ET_0 estimates and 12 meteorological variables in Iran using the trend-free pre-whitening Mann–Kendall (TFPW-MK) and Spearman's Rho tests. These sites were chosen to cover four different climates, namely arid, semiarid, Mediterranean and humid. Moreover, variations of 12 meteorological variables (mean, maximum, and minimum air temperature, difference between the maximum and minimum air temperature, rainfall, wind speed and direction, mean and minimum relative humidity, vapor pressure deficit, number of cloudy days, and sunshine hours) were investigated to characterize forces that drive trends of ET_0. Furthermore, the effect of serial correlations on the trends was eliminated by the TFPW approach [24]. Finally, during the period 1961–2010, the coincidence of annual extremum values of ET_0 and meteorological variables was studied to identify the main meteorological variables that affect variations of ET_0.

2. Data

Daily meteorological data in the 18 study sites were downloaded from the Iran Meteorological Organization (IMO) archive. These data cover a period of 50 years (1961–2010), and include mean, minimum, and maximum daily air temperature (°C), vapor pressure deficit (kPa), mean and minimum relative humidity (%), wind speed (m/s) at the screen-height of 2 m, rainfall (mm/day), number of cloudy days, and number of sunshine hours per day (hr/day).

Characteristics of the study sites are indicated in Table 1. Figure 1 shows the location of these sites in Iran. They are chosen to cover various climate regions (arid, semiarid, humid, and Mediterranean) across Iran.

Table 1. Location, altitude, and climate of the 18 study sites.

Synoptic Station	ICAO Code	Latitude (°N)	Longitude (°E)	Altitude (masl)	Climate Type
Ahvaz (AH)	40811	31°20′	48°40′	22.5	Arid
Arak (AR)	40769	34°6′	49°46′	1708.0	Semiarid
Bushehr (BU)	40858	28°58′	50°49′	9.0	Arid
Esfahan (ES)	40800	32°37′	51°40′	1550.4	Arid
Hamedan (HA)	40768	34°52′	48°32′	1741.5	Semiarid
Jiroft (JI)	40866	28°35′	57°48′	601.0	Arid
Kerman (KE)	40841	30°15′	56°58′	1753.8	Arid
Mashhad (MA)	40745	36°16′	59°38′	999.2	Semiarid
Moghan (MO)	40700	39°39′	47°55′	31.9	Semiarid
Qazvin (QA)	40731	36°15′	50°3′	1279.2	Semiarid
Rasht (RA)	40719	37°19′	49°37′	−8.6	Humid
Sanandaj (SA)	40747	35°20′	47°0′	1373.4	Mediterranean
Shahrekord (SK)	40798	32°17′	50°51′	2048.9	Semiarid
Shiraz (SH)	40848	29°32′	52°36′	1484.0	Semiarid
Tabriz (TA)	40706	38°5′	46°17′	1361.0	Semiarid
Urmia (UR)	40712	37°40′	45°3′	1328.0	Semiarid
Yazd (YA)	40821	31°54′	54°17′	1237.2	Arid
Zabol (ZA)	40829	31°2′	61°29′	489.2	Arid

ICAO: International Civil Aviation Organization; masl: meter above sea level.

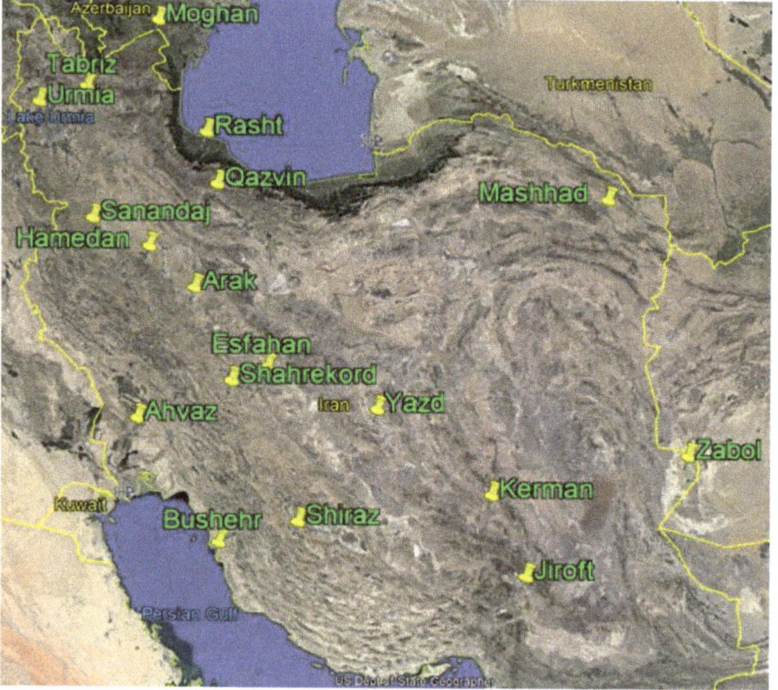

Figure 1. Location of the 18 study sites in Iran.

3. Models and Methods

3.1. FPM Equation

ET_0 is the rate of evapotranspiration from a uniform height, actively growing, well-watered, and completely shaded hypothetical crop [33]. The hypothetical crop has a height 0.12 (m), a fixed surface resistance of 70 (s/m), and an albedo of 0.23, closely resembling the evapotranspiration from an extensive surface of green grass [26]. In this study, daily meteorological variables from the 18 study sites were used in the Agricultural Organization of the United Nation (FAO)-Penman–Monteith (FPM) equation to estimate ET_0 on a daily basis. ET_0 estimates from the FPM equation were validated against the lysimeter measurements in the 18 study sites [34–37].

The FPM equation is given by [33]:

$$ET_0 = \frac{0.408(R_n - G) + \gamma \frac{900}{T+273} u(es - ea)}{\Delta + \gamma(1 + 0.34u)} \tag{1}$$

where ET_0 is the daily reference evapotranspiration (mm/day), γ is the psychrometric constant (kPa/°C), Δ is the slope of the saturation vapor pressure–temperature curve (kPa/°C), T is the mean daily air temperature (°C), and u is the mean daily wind speed at the screen-height of 2 m (m/s). G is the ground heat flux (MJ/m^2/day) and is negligible on daily timescales [21,33].

es and ea are, respectively, the saturation and actual vapor pressures (kPa), which are given by [33]:

$$es = \frac{e_{(Tmax)} + e_{(Tmin)}}{2} \tag{2}$$

$$ea = \frac{e_{(Tmin)} \frac{RHmax}{100} + e_{(Tmax)} \frac{RHmin}{100}}{2} \tag{3}$$

where $e_{(Tmax)}$ and $e_{(Tmin)}$ are the saturation vapor pressures (kPa) at daily maximum and minimum air temperatures, respectively. *RHmax* and *RHmin* are the daily maximum and minimum relative humidity (%), respectively.

R_n is the net radiation (MJ/m^2/day), and is estimated by [21,33]:

$$R_n = \left(a_s + b_s \frac{n}{n_{max}}\right) R_a \tag{4}$$

where n is the number of sunshine hours per day (hr/day), n_{max} is the maximum possible duration of sunshine per day (hr/day), and R_a is the extraterrestrial radiation (MJ/m^2/day). R_a and n_{max} depend on the latitude and julian day [33]. a_s and b_s are empirical coefficients, which are obtained from [21] for each study site.

In this study, the annual averages of daily FPM ET_o estimates and meteorological variables were used to analyze the annual trends.

3.2. Mann–Kendall (MK) Test

One of the most well-known methods for detecting trends in a time series is the non-parametric Mann–Kendall (MK) test [38–40]. Unlike the parametric statistical approaches, the non-parametric statistical tests are more suitable for non-Gaussian distributed data, which are frequently observed in hydrologic time series [34]. The MK test is defined as follows [38–40]:

$$S = \sum_{i=1}^{N-1} \sum_{j=i+1}^{N} sign\left(x_j - x_i\right) \tag{5}$$

$$sign\left(x_j - x_i\right) = \begin{cases} 1 \; if \; \left(x_j - x_i\right) > 0 \\ 0 \; if \; \left(x_j - x_i\right) = 0 \\ -1 \; if \; \left(x_j - x_i\right) < 0 \end{cases} \tag{6}$$

$$VAR(S) = \frac{1}{18}\left[N(N-1)(2N+5) - \sum_{p-1}^{g} t_p\left(t_p - 1\right)\left(2t_p + 5\right)\right] \tag{7}$$

$$If \; S > 0 \; then \; Z = (S - 1)/VAR \; (S)^{0.5} \tag{8a}$$

$$If \; S = 0 \; then \; Z = 0 \tag{8b}$$

$$If \; S < 0 \; then \; Z = (S + 1)/VAR(S)^{0.5} \tag{8c}$$

where N is the number of data, S is the summation of signs, VAR is the variance, and x_j and x_i are the data values in years j and i, respectively (with $j > i$). t_p and g indicate ties and the number of ties, respectively. The MK test determines whether to reject the null hypothesis (H0) or accept the alternative hypothesis (Ha), where H0: no monotonic trend is present and Ha: a monotonic trend is present. If $1.65 < Z \le 1.96$, $1.96 < Z \le 2.58$, $Z > 2.58$, the trend is significant at the confidence levels of 90%, 95%, and 99%, respectively [38–40].

Local (at-site) significance levels for each trend test can be obtained from

$$p = 2[1 - \Phi(|Z|)] \tag{9}$$

where $|\ |$ denotes the absolute value, and Φ ( ) is defined as,

$$\Phi(|Z|) = \frac{1}{\sqrt{2\pi}} \int_0^{|Z|} \exp\left(\frac{t^2\theta}{2}\right) dt \tag{10}$$

where θ is the random variable, and t is the variable for which the cumulative distribution function should be calculated. If $p \leq \alpha$, the existing trend is statistically significant at the significance level of α [41,42].

Sometimes the MK test detects trends because of the serial correlations of time series data, which leads to an increased rejection rate of the null hypothesis [41,43].

Hydrologic time series often have significant serial correlations that can undermine the ability of the MK test to correctly assess the significance of trend [41,43].

In this study, trend-free pre-whitening (TFPW) was used to effectively eliminate the impact of serial correlations on the MK test [41,43–45].

This method is defined by the following steps:

1. Calculate the slope of trend using the Sen's slope estimator [43,44]:

$$Q_i = \frac{x_j - x_k}{j - k} \tag{11}$$

$$Q = \begin{cases} Q_{\frac{N+1}{2}} & N \text{ is odd} \\ \frac{1}{2}\left(Q_{\frac{N}{2}} + Q_{\frac{N+2}{2}}\right) & N \text{ is even} \end{cases} \tag{12}$$

where Q stands for the slope of trend. Q_i is the Sen's slope estimator for each value of i, x_j and x_k are the numerical values at times j and k ($j > k$), respectively.

2. There is no trend if Q is equal to zero. Otherwise, it is assumed that the existing trend is monotonic, and the time series data are de-trended as follows:

$$X_t' = X_t - Qt \tag{13}$$

where X_t' is the lag-1 autocorrelation of the de-trended time series and X_t is the autocorrelation of time series.

3. Using the rank correlation coefficient estimator, the lag-1 autocorrelation of the de-trended time series is estimated by replacing the sample data by their ranks as follows [45]:

$$r_j = \frac{\frac{1}{n-j}\sum_{i=1}^N \left(X_i - \overline{X}\right)\left(X_{i+j} - \overline{X}\right)}{\frac{1}{n}\sum_{i=1}^N \left(X_{i+j} - \overline{X}\right)^2} \tag{14}$$

where r_j is the lag-j autocorrelation coefficient, and X is the average of the data. Then, the estimated lag-1 autocorrelation is removed from the time series as follows:

$$X_t'' = X_t' - r_j X_{t-1}' \tag{15}$$

4. Adding the removed trend in step 2:

$$X_t''' = X_t'' + Qt \tag{16}$$

3.3. Spearman's Rho Test

The Spearman's Rho test was applied to investigate correlation among all the meteorological variables and ET_0 estimates from the FPM equation.

The Spearman's Rho test is defined as [18]:

$$\rho = 1 - \frac{6 \sum_{i=1}^{N} (x_i - i)^2}{N(N^2 - 1)} \tag{17}$$

$$Z = \rho \sqrt{N - 1} \tag{18}$$

where ρ is the correlation coefficient of linear regression between series i (the order of the data in the original series) and x (data values) [18]. If $|Z| > Z_\alpha$ at a significance level of α, then the null hypothesis of no trend is rejected [18].

Following the literature [1–24], the confidence levels of 90%, 95%, and 99% were used in this study to evaluate the trends of ET_0 and meteorological variables.

If there is no mention of confidence level, it meant a confidence level of 95% was adopted. Otherwise, we explicitly mention the confidence levels of 90% and 99% wherever they were used.

4. Results and Discussions

4.1. Comparison of p-Values from the MK and TFPW-MK Tests

Table 2 compares p-values from the MK and TFPW-MK tests for ET_0 and the 12 meteorological variables in the 18 study sites.

The p-value is a random variable derived from the distribution of the test statistic to analyze a data set and to test a null hypothesis [41,42]. If p-values from the abovementioned tests are different, but they result in an identical trend for a particular time series (i.e., an insignificant trend or a significant trend at the same confidence level), the cells in Table 2 are highlighted in yellow color. If the two tests yield different trends for a specific time series (i.e., one leads to a significant trend, while the other one results in an insignificant trend, or they both lead to a significant trend but with different confidence levels), the cells are highlighted in red color.

As shown, in each study site, p-values from the MK and TFPW-MK tests are different at least for three meteorological variables (yellow and red cells). Overall, 38% of the obtained p-values from the two tests are different due to the serial correlations of time series data. In each study site, the two tests also lead to different trends for at least one variable (red cells). As shown in Table 2, 15% of the trends from two tests are different. These results indicate that the serial correlation undermine the ability of the MK test to correctly determine the trends and their confidence level [41,43]. Hence, the TFPW method should be applied to eliminate the impact of the serial correlations on the MK test.

As can be seen in Table 2, some meteorological variables are subject to different p-values from the MK and TFPW-MK test more often than others. For example, wind direction (*WD*), wind speed (*WS*), and rainfall (*P*) were flagged for more study sites compared to mean air temperature (*Tmean*) and minimum air temperature (*Tmin*). This happens because *WD*, *WS*, and *P* time series have more serial correlations compared to *Tmean* and *Tmin* time series. Similar findings were reported in other studies [46–49].

Table 2. Estimated *p*-values from the Mann–Kendall (MK) and trend-free pre-whitening Mann–Kendall (TFPW-MK) tests in the 18 study sites.

Sites	Method	ET_o	Tmean	Tmin	Tmax	Tmax-Tmin	es-ea	RH	RHmin	P	WD	WS	CD	n
Ahvaz	TFPW-MK	0.146	0.000	0.000	0.001	0.000	0.170	0.725	0.000	0.598	0.430	0.907	0.040	0.041
	MK	0.276	0.000	0.000	0.001	0.000	0.059	0.725	0.000	0.598	0.061	0.928	0.003	0.101
Arak	TFPW-MK	0.003	0.682	0.358	0.980	0.255	0.332	0.645	0.462	0.066	0.101	0.034	0.353	0.719
	MK	0.000	0.682	0.358	0.981	0.255	0.332	0.632	0.462	0.066	0.581	0.034	0.420	0.464
Bushehr	TFPW-MK	0.184	0.000	0.000	0.002	0.006	0.031	0.238	0.270	0.340	0.121	0.003	0.146	0.011
	MK	0.065	0.000	0.000	0.007	0.006	0.031	0.349	0.270	0.301	0.002	0.003	0.041	0.011
Esfahan	TFPW-MK	0.000	0.000	0.034	0.000	0.153	0.001	0.155	0.058	0.063	0.828	0.000	0.001	0.096
	MK	0.000	0.000	0.034	0.000	0.153	0.001	0.155	0.058	0.013	0.944	0.000	0.001	0.005
Hamedan	TFPW-MK	0.987	0.019	0.014	0.358	0.143	0.980	0.024	0.110	0.732	0.920	0.366	0.017	0.137
	MK	0.987	0.019	0.014	0.414	0.143	0.980	0.024	0.048	0.732	0.965	0.366	0.017	0.137
Jiroft	TFPW-MK	0.037	0.291	0.717	0.025	0.003	0.085	0.131	0.027	0.004	0.608	0.349	0.110	0.566
	MK	0.037	0.291	0.717	0.025	0.003	0.085	0.131	0.027	0.002	0.769	0.008	0.110	0.566
Kerman	TFPW-MK	0.032	0.000	0.000	0.000	0.047	0.412	0.622	0.375	0.039	0.119	0.014	0.000	0.034
	MK	0.000	0.000	0.000	0.000	0.047	0.412	0.193	0.475	0.001	0.001	0.014	0.000	0.034
Mashhad	TFPW-MK	0.001	0.000	0.000	0.002	0.000	0.001	0.020	0.040	0.947	0.043	0.128	0.744	0.472
	MK	0.001	0.000	0.000	0.002	0.000	0.005	0.020	0.040	0.947	0.288	0.085	0.744	0.277
Moghan	TFPW-MK	0.359	0.004	0.000	0.113	0.574	0.441	0.091	0.574	0.692	0.278	0.348	0.032	0.441
	MK	0.359	0.004	0.000	0.225	0.574	0.441	0.091	0.539	0.692	0.058	0.348	0.032	0.441
Qazvin	TFPW-MK	0.000	0.362	0.457	0.417	0.394	0.000	0.001	0.035	0.610	0.316	0.000	0.001	0.847
	MK	0.000	0.362	0.457	0.417	0.394	0.000	0.000	0.035	0.610	0.038	0.000	0.001	0.803
Rasht	TFPW-MK	0.375	0.000	0.000	0.558	0.000	0.003	0.001	0.340	0.598	0.472	0.569	0.452	0.328
	MK	0.097	0.000	0.000	0.351	0.000	0.001	0.000	0.445	0.598	0.803	0.569	0.262	0.328
Sanandaj	TFPW-MK	0.358	0.000	0.047	0.000	0.120	0.013	0.124	0.000	0.011	0.178	0.913	0.001	0.046
	MK	0.161	0.000	0.052	0.000	0.198	0.013	0.119	0.000	0.011	0.589	0.727	0.000	0.090
Shahrekord	TFPW-MK	0.000	0.001	0.000	0.195	0.091	0.000	0.106	0.000	0.757	0.389	0.000	0.094	0.320
	MK	0.000	0.001	0.002	0.195	0.091	0.000	0.106	0.000	0.610	0.061	0.000	0.175	0.320
Shiraz	TFPW-MK	0.024	0.000	0.000	0.001	0.001	0.000	0.037	0.012	1.000	0.821	0.000	0.738	0.457
	MK	0.024	0.000	0.000	0.001	0.001	0.000	0.037	0.002	1.000	0.933	0.000	0.738	0.415
Tabriz	TFPW-MK	0.225	0.000	0.000	0.000	0.744	0.000	0.000	0.000	0.000	0.682	0.634	0.000	0.213
	MK	0.193	0.000	0.000	0.000	0.744	0.000	0.000	0.000	0.006	0.864	0.634	0.000	0.091
Urmia	TFPW-MK	0.015	0.069	0.061	0.022	0.375	0.320	0.375	0.000	0.153	0.130	0.021	0.002	0.049
	MK	0.006	0.054	0.061	0.022	0.375	0.076	0.490	0.000	0.107	0.367	0.021	0.002	0.014
Yazd	TFPW-MK	0.763	0.000	0.000	0.000	0.000	0.001	0.076	0.040	0.384	0.206	0.050	0.068	0.019
	MK	0.679	0.000	0.000	0.000	0.000	0.000	0.149	0.185	0.384	0.093	0.013	0.112	0.009
Zabol	TFPW-MK	0.000	0.000	0.000	0.011	0.751	0.201	0.349	0.007	0.403	0.101	0.000	0.462	0.763
	MK	0.000	0.002	0.000	0.000	0.775	0.201	0.460	0.042	0.001	0.101	0.000	0.386	0.763

4.2. Trends of ET_0 and Meteorological Variables in the Study Sites

Variations of the meteorological variables were assessed in all of the study sites to identify the ones that control the trend of annual mean of ET_0. However, herein, the trends are presented only in two sites, Kerman and Qazvin (Figures 2 and 3).

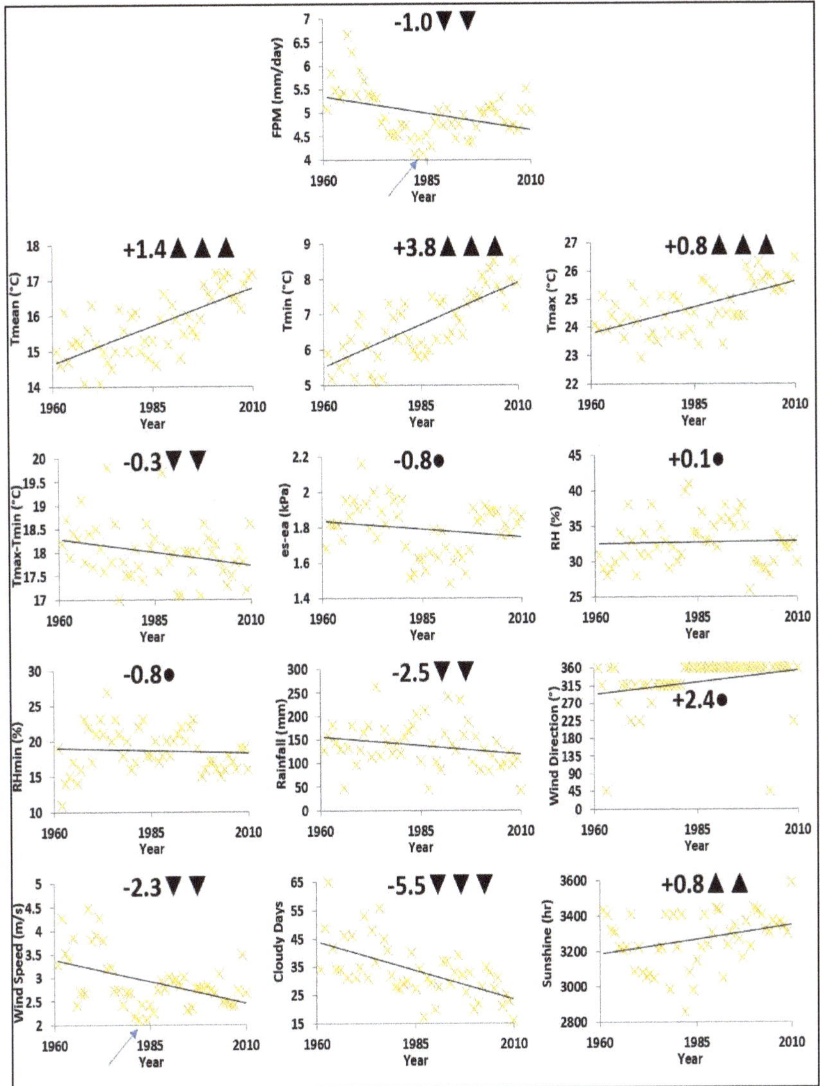

Figure 2. The 50-year (1960–2010) trends of annually averaged reference evapotranspiration and meteorological variables in Kerman. Circles indicate an insignificant trend. One, two, and three triangles indicate a significant trend at a confidence level of 90%, 95%, and 99%, respectively. The value on the left of triangular symbols shows the increasing/decreasing rate over the 50-year period.

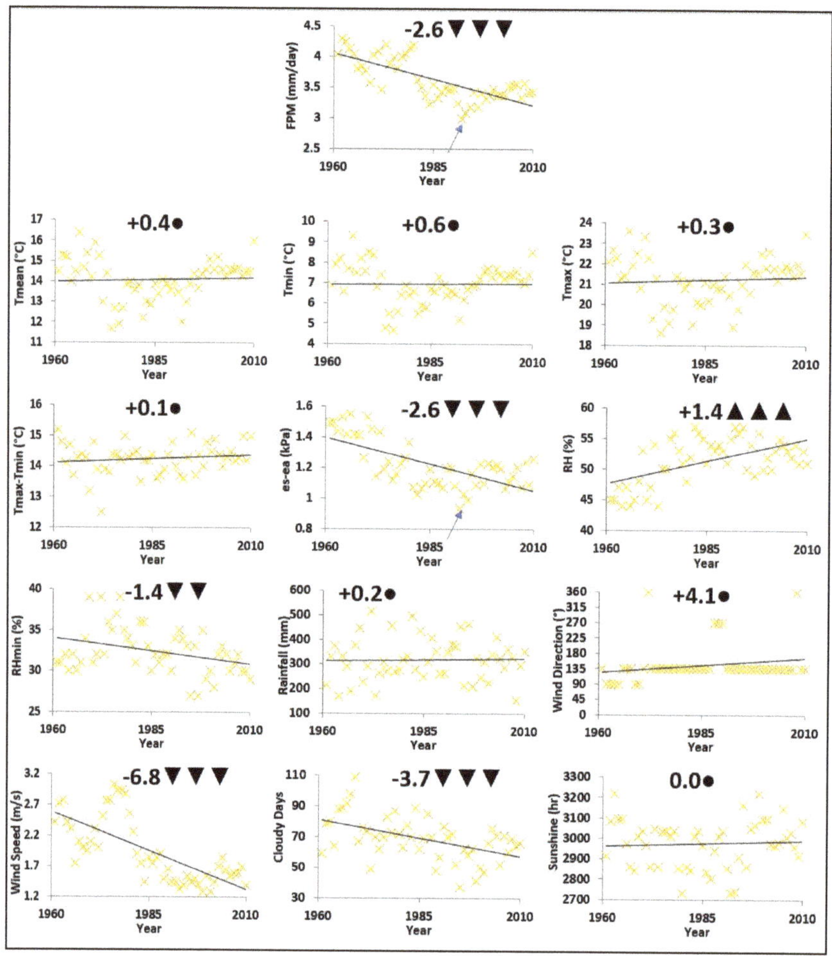

Figure 3. The 50-year (1960–2010) trends of annually averaged reference evapotranspiration and meteorological variables in Qazvin. Circles indicate an insignificant trend. One, two, and three triangles indicate a significant trend at a confidence level of 90%, 95%, and 99%, respectively. The value on the left of triangular symbols shows the increasing/decreasing rate over the 50-year period.

4.2.1. Kerman

Figure 2 shows the 50-year trends in the annual averages of daily FPM ET_0 estimates, and meteorological variables including mean (*Tmean*), minimum (*Tmin*), and maximum (*Tmax*) air temperature, the difference between maximum and minimum air temperature (*Tmax-Tmin*), vapor pressure deficit (*es-ea*), relative humidity (*RH*), minimum relative humidity (*RHmin*), rainfall (*P*), wind speed and direction (*WS* and *WD*), number of cloudy days (*CD*), and sunshine hours (*n*) in Kerman.

As can be seen, the significant upward trends in *Tmean*, *Tmin*, *Tmax*, and *n* (at confidence level 95%) as well as the substantial downward trends in *CD*, *Tmax-Tmin*, *P*, and *WS* (at confidence level 95%) led to a negative trend (at confidence level 95%) in the FPM ET_0 estimates.

Although the upward trends in *Tmax*, *Tmin*, *Tmean*, and *n* could potentially increase ET_0, a downward trend in *WS* and (*es-ea*) led to a significant decreasing trend in the FPM ET_0 values. It is worth mentioning that the minimum of both *WS* and ET_0 occurred in the same years (i.e., 1982 and

1984) (see the arrows in Figure 2). Similarly, the course of (*es-ea*) shows lower values in 1982–1986. Thus, *WS* and to a lesser extent (*es-ea*) may be the driving force for the variations of ET_0 in Kerman.

Although *CD* showed the strongest downward trend among all the meteorological variables in Kerman, it could not capture the variations of ET_0 as good as *WS*. Both *WS* and ET_0 showed significant trends at a confidence level of 95%, while *CD* showed a significant trend at a confidence level of 99%.

During 1960–1980s, the north-western wind was prevailing, which blows from the Zangi Abad region with an arid climate into Kerman [50,51]. However, from 1980 until 2010, the northern wind was dominant, which comes from Chatroud region with dry temperate climate [50,51]. Hence, the change in wind direction (*WD*) reduced the impact of the arid climate in Kerman, and consequently decreased ET_0.

The decreasing rate of *Tmax-Tmin* (−0.3%/decade) was due to a more rapid increase in *Tmin* (+3.8%/decade) compared to *Tmax* (+0.8%/decade). *Tmin*, *P*, *WS*, and *CD* had the highest changes (larger than ±2% per decade), implying climate change in Kerman [50,51]. The lowest *P* and *CD*, and the highest *Tmax* and *n* values (during the 50-year period) were observed in 2010, which may an indication of drought in this region [50,52–57]. These results show the complexity of forces driving the trend of ET_0 and may highlight importance of considering different ET_0 models and variables in various climates for local and global studies.

4.2.2. Qazvin

Figure 3 shows the 50-year trends of annually averaged ET_0 and meteorological variables in Qazvin. As can be seen in Figure 3, the significant upward trend in *RH* as well as substantial downward trends in *es-ea*, *WS*, *CD*, and *RHmin* (at the confidence level 95%) led to a negative trend in the FPM ET_0 estimates. It is worth mentioning that these significant trends reduced ET_0 and may cause alarming climate change in Qazvin [57–59]. During the 50-year study period (1961–2010), *Tmax* reached its highest value in 2010. It should be noted that the minimum *es-ea* and ET_0 occurred in 1991 (see the arrows in Figure 3). The highest decreasing rate was for *WS* (−6.8%/decade). *RH* showed an upward trend as a result of decreasing *WS* and *es-ea* (Figure 3). The warm dry south-eastern winds (called Raz or Shareh) originated from the arid areas of central Iran (Yazd). They were prevailing from 1961 to 2010 and decreased *RHmin* in Qazvin [60–63].

4.3. Trends of ET_0 and Meteorological Variables in Iran

Table 3 and Figure 4 summarize trends of FPM ET_0 estimates and meteorological variables in the 18 study sites. The FPM ET_0 estimates showed significant upward and downward trends, respectively, in 33.3% (28.6% of arid regions vs. 44.4% of semiarid regions) and 22.2% (28.6% of arid regions vs. 22.2% of semiarid regions) of the study sites (Table 3 and Figure 4). Hence, the FPM ET_0 retrievals had significant variations in most parts of Iran (57.2% of arid regions vs. 66.6% of semiarid regions), which may indicate climatic change signals.

Table 3. Trends of Agricultural Organization of the United Nation (FAO)-Penman–Monteith (FPM) evapotranspiration (ET_o) estimates and 12 meteorological variables in the 18 study sites. U and D represent the significant upward and downward trends, respectively. NS denotes no significant trend.

Variations	ET_o	Tmean	Tmin	Tmax	Tmax-Tmin	es-ea	RH	RHmin	P	WD	WS	CD	n
Ahvaz	NS	U ***	U ***	U ***	D ***	NS	NS	D ***	NS	NS	NS	U **	U **
Arak	U ***	NS	NS	NS	NS	NS	NS	NS	D *	NS	U **	NS	NS
Bushehr	NS	U ***	U ***	U ***	D ***	U **	NS	NS	NS	NS	D ***	NS	D **
Esfahan	D ***	U ***	U **	U ***	NS	U ***	NS	D **	U *	NS	D ***	D ***	U *
Hamedan	NS	U **	U **	NS	NS	NS	U **	NS	NS	NS	NS	U **	NS
Jiroft	U **	U **	NS	U **	U ***	U *	NS	D **	D ***	NS	NS	NS	NS
Kerman	D **	U ***	U ***	U ***	D **	NS	NS	NS	D **	NS	D **	D ***	U **
Mashhad	U ***	U ***	U **	U ***	D ***	U ***	D **	D **	NS	D **	NS	NS	NS
Moghan	NS	U ***	U ***	NS	NS	NS	U *	NS	NS	NS	NS	D **	NS
Qazvin	D ***	NS	NS	NS	NS	D ***	U ***	D **	NS	NS	D ***	D ***	NS
Rasht	NS	U ***	U ***	NS	D ***	D ***	U ***	NS	NS	NS	NS	NS	NS
Sanandaj	NS	U ***	U **	U ***	NS	D **	NS	D ***	D **	NS	NS	D ***	U **
Shahrekord	U ***	D ***	D ***	NS	U *	D ***	NS	D ***	NS	NS	U ***	D *	NS
Shiraz	D **	U ***	U ***	U ***	D ***	U ***	D **	D **	NS	NS	D ***	NS	NS
Tabriz	NS	U ***	U ***	U ***	NS	U ***	D ***	D ***	D ***	NS	NS	D ***	NS
Urmia	U **	U **	U *	U **	NS	NS	NS	D ***	NS	NS	U *	D ***	U *
Yazd	NS	U ***	U ***	U ***	D ***	U ***	D *	D **	NS	NS	D **	U *	U **
Zabol	U ***	U ***	U ***	U **	NS	NS	NS	D ***	NS	NS	U ***	NS	NS
U (%)	33.3	77.8	77.8	66.7	11.1	38.9	22.2	0.0	5.6	0.0	22.2	16.7	33.3
D (%)	22.2	5.6	5.6	0.0	38.9	22.2	22.2	66.7	27.8	5.6	33.3	44.4	5.6
U+D (%)	55.5	83.4	83.4	66.7	50.0	61.1	44.4	66.7	33.4	5.6	55.5	61.1	38.9

*, **, *** are significant trends at a confidence level of 90%, 95%, and 99%, respectively.

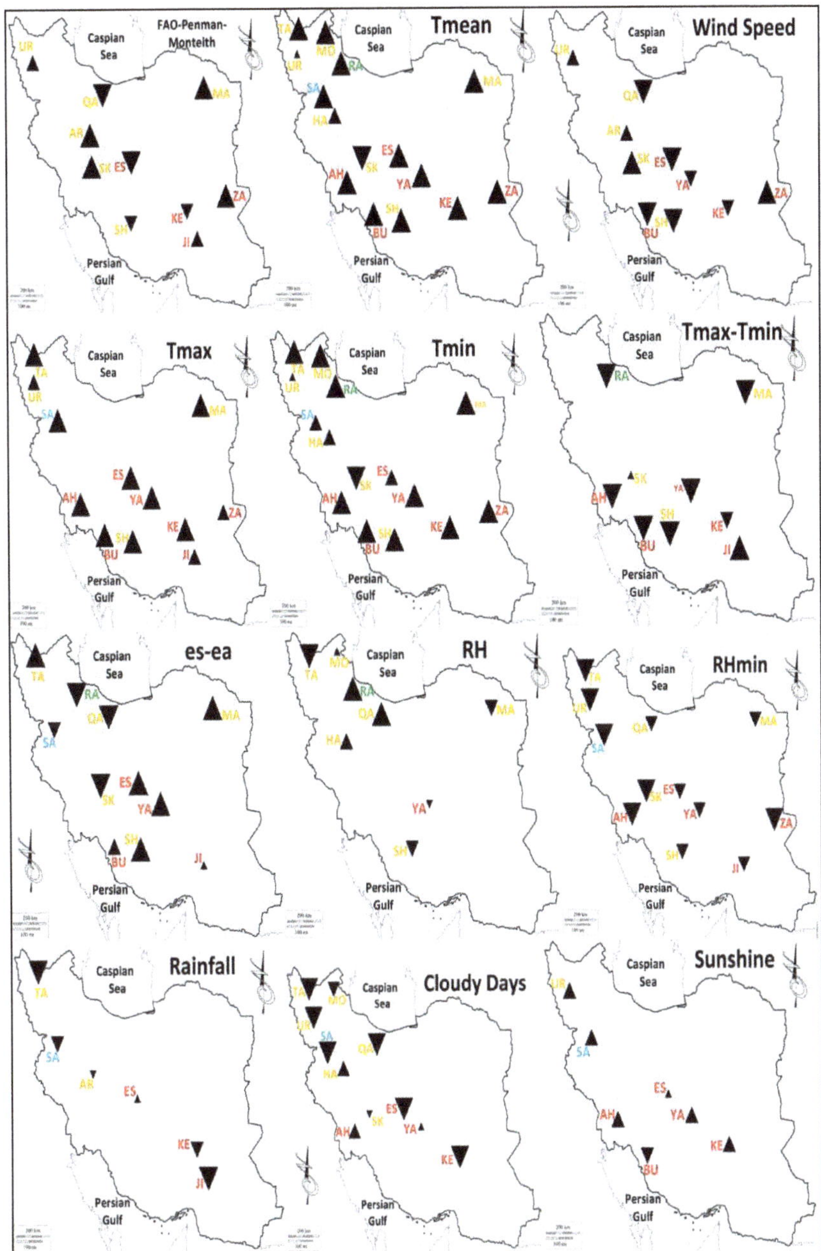

Figure 4. The average of variations in reference evapotranspiration and meteorological variables in Iran from 1961 to 2010. Big, medium, and small triangles imply significant values at a confidence level of 99%, 95%, and 90%, respectively. Blue, red, yellow, and green colors indicate Mediterranean, arid, semiarid and humid climates, respectively.

Results showed that ET_0 had a significant positive (negative) trend in the west and east (center) of Iran. *WS* was the only meteorological variable with the same rising/falling trend.

As shown in Table 3, in eight of the study sites (i.e., 44.4% of sites), there were significant trends in both ET_0 and $Tmean/RHmin/WS$. Moreover, both ET_0 and $Tmin/Tmax/es-ea$ showed significant trends in 33.3% of the study sites. Hence, these meteorological variables could be considered as driving forces to control variations of ET_0 in Iran.

WS had significant upward and downward trends in 22.2% (14.3% of arid regions vs. 33.3% of semiarid regions) and 33.3% (57.1% of arid regions vs. 22.2% of semiarid regions) of the study sites, respectively. Therefore, significant variations in WS were observed in 55.5% (71.4% of arid regions vs. 55.5% of semiarid regions) of the sites. Table 3 and Figures 2–4 indicate that WS is the driving force for the trends of ET_0 in 55.5% of the study sites. Thus, WS may be the most important variable that controls the variations of ET_0, particularly in the arid regions. These findings are further validated by our outcomes in Table 5 (Section 4.6) that lists the three meteorological variables with the highest correlation with ET_0. As shown, WS has the highest significant correlation with ET_0 in all the study sites expect Rasht (RA). Studying the trends of ET_0 and meteorological variables over longer time periods and more sites across Iran leads to more reliable results.

Table 3 and Figure 4 also showed that there were significant upward trends for $Tmean$ and $Tmax$ in 77.8% (85.7% of arid regions vs. 55.6% of semiarid regions) and 66.7% (100.0% of arid regions vs. 33.3% of semiarid regions) of the study sites, respectively. In Jiroft, Mashhad, and Urmia (3 out of 18 study sites, i.e., 16.7% study sites), there were upward significant trends between ET_0 and $Tmean/Tmax$ (Table 3). There was a significant downward trend for $RHmin$ in 66.7% (71.4% of arid regions vs. 66.7% of semiarid regions) of the study sites (Figure 4).

$Tmax$, $Tmean$, $Tmin$, $es-ea$, and n had significant increasing trends, respectively, in 100.0%, 85.7%, 85.7%, 57.1%, and 57.1% of the arid sites. On the other hand, $RHmin$, $Tmax-Tmin$, and WS had significant decreasing trends in 71.4%, 57.1%, and 57.1% of the arid sites, respectively. These results indicated that the arid sites were affected more than the semiarid ones by climate change. In most of the semiarid sites, $Tmin$ and $Tmean$ showed increasing, and $RHmin$ and CD indicated decreasing trends.

According to the results, P, WS, and CD showed significant trends in 42.9%, 28.6%, and 28.6% of the arid sites, respectively. WS, $Tmin$, $RHmin$, P, and CD had substantial trends in 44.4%, 22.2%, 11.1%, 11.1%, and 11.1% of the semiarid sites, respectively. Furthermore, WS, $Tmin$, and CD had a rate of more than $\pm2\%$/decade in 66.7% (85.7% of arid regions vs. 66.7% of semiarid regions), 66.7% (42.9% of arid regions vs. 77.8% of semiarid regions), and 55.6% (57.1% of arid regions vs. 55.6% of semiarid regions) of the study sites, respectively.

As mentioned above, variations of ET_0 in each study site are controlled by meteorological variables such as air temperature and wind speed. The Arak and Esfahan sites show contrasting trends in ET_0 although they are relatively close to each other. This happens because wind speed (which is the dominant controlling variable in both sites) has an increasing (decreasing) trend in Arak (Esfahan). Furthermore, Arak and Esfahan have different climate types. Arak is located in a cold mountainous area, while Esfahan is placed in a hot flat desert.

The focus of this study was to evaluate the annual trends of ET_0 and meteorological variables over a 50-year period (1961–2010). However, according to Table 3 and Figure 4, we can conclude that their seasonality is important. For example, the annual minimum air temperature, $Tmin$ (that occurs in cold seasons, i.e., fall–winter) had significant increasing trends in 83.4% of sites. While, the annual maximum air temperature, $Tmax$ (which happens in warm seasons, i.e., spring–summer) showed significant rising trends in 66.7% of sites. In addition, the significant downward trends of $Tmax-Tmin$ were observed in 41.5% of sites, whereas its upward trends were seen only in 9.3% of sites. Its significant downward trends were due to the higher increasing rate of $Tmin$ in cold seasons than that of $Tmax$ in warm seasons. These results imply that the cold seasons have more influence on the significant trends than warm seasons. It should be noted that these findings are primitive, and future studies should be directed towards evaluating the seasonal variations of ET_0 and other climatic variables over long periods.

4.4. Range of Variations of ET_0 and Meteorological Variables

Figure 5 shows the box plots for the 50-year variations of ET_0 and meteorological variables in each study site.

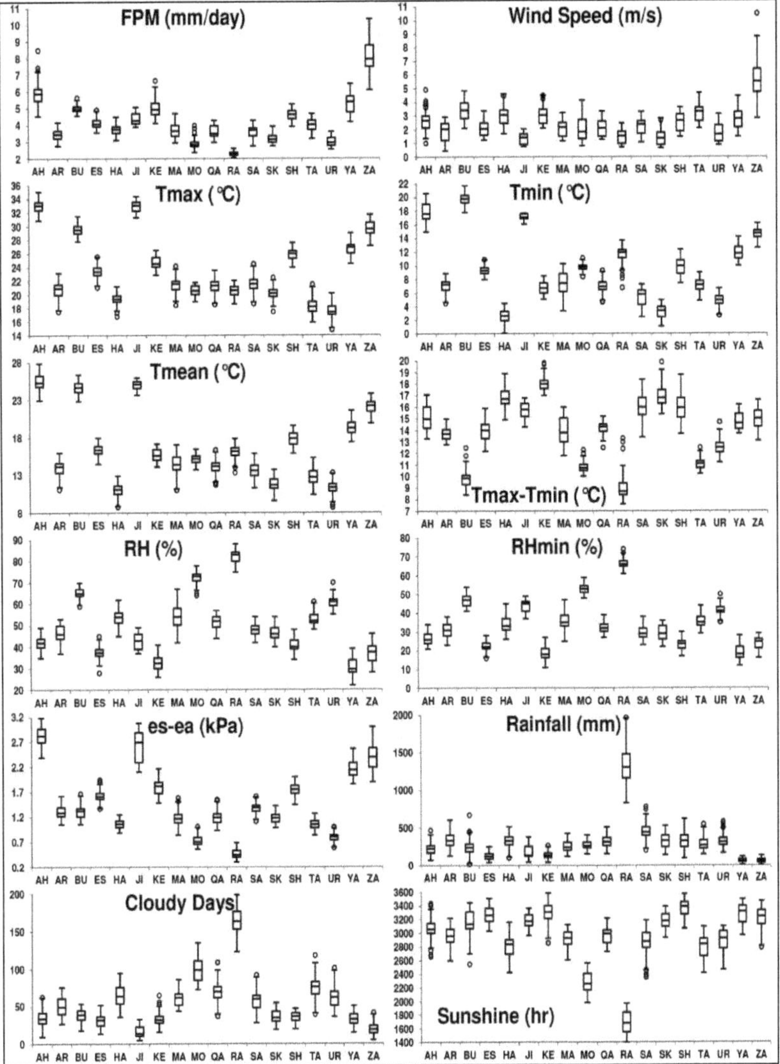

Figure 5. Statistical boxplots of meteorological variables and reference evapotranspiration in each study site. For each distribution, the horizontal line within the box indicates the median (50% percentile). The upper and lower edges of the box represent the 75% and 25% percentile, respectively. The upper and lower ends of the whiskers represent the maximum and minimum values. Outliers are observations beyond the end of whiskers.

As can be seen, the largest 50-year variations in both ET_0 and *WS* happened in Zabol (ZA), which highlighted the significant influence of *WS* on ET_0.

The second highest variations of ET_0 was seen in Ahvaz (AH). AH had also the highest variations of *Tmean* compared to the other study sites. Hence, the variations of ET_0 in Ahvaz can be related to that of *Tmean*.

The lowest variations in *Tmax-Tmin*, *es-ea*, and *n* values were observed in Rasht (RA), which led to the smallest 50-year change in ET_0 [9,10]. Most of the outliers occurred in 2010, which signaled a drought condition in different regions of Iran. The FPM ET_0 values ranged from 2 to 11 mm day^{-1} in various climates of Iran.

As shown in Equation (1), ET_0 is affected by the atmospheric vapor pressure deficit (*es-ea*). As expected, a high positive correlation exists between ET_0 and *es-ea* in Figure 5. For example, high values of ET_0 and *es-ea* are seen in Zabol (ZA). Moreover, the lowest ET_0 and *es-ea* values are observed in Rasht (RA). This happens because the atmospheric demand to water vapor increases as the vapor pressure deficit (*es-ea*) rises [33]. Positive correlations are also observed between ET_0 and *Tmax*, and ET_0 and *n*. A larger *Tmax* leads to more potential for converting soil moisture to water vapor and causes plants to open up their stomata and release more water vapor [33]. Furthermore, larger values of *n* yield higher R_n (Equation (4)) and ET_0 (Equation (1)). On the other hand, a negative correlation is found between ET_0 and *RH*. The highest ET_0 and lowest *RH* values are seen in RA. This is due to the fact that the atmospheric demand for water vapor decreases as *RH* increases.

4.5. Annual Extremum Values of ET_0 and Meteorological Variables during the Study Period (1961–2010)

Table 4 shows the years of occurrence of maximum and minimum values of ET_0 and meteorological variables in each study site during the 50-year study period (i.e., 1961–2010).

Table 4. The years of occurrence of maximum and minimum values of ET_0 and meteorological variables in each study site during the 50-year study period. If the annual extremum values of ET_0 and a particular meteorological variable coincide, they are highlighted in green color.

Sites		ET_o	Tmean	Tmin	Tmax	Tmax-Tmin	es-ea	RH	RHmin	P	WS	CD	n
Ahvaz	Max	1964	2010	2010	2010	1973	1973	1982	1961	1997	1964	1982	1998
(arid)	Min	1962	1969	1968	1992	1992	1984	1964	1990	1973	1979	1964	1992
Arak	Max	1973	1966	1966	2010	1964	1973	1992	1993	1969	1982	1974	1973
(semiarid)	Min	1966	1992	1983	1992	1996	1992	1973	2010	1973	1966	2010	1984
Bushehr	Max	1967	2010	2010	1962	1962	1981	1972	1963	1997	1967	1974	1970
(arid)	Min	1979	1964	1964	1972	2009	1972	1981	1981	2010	1998	2008	1992
Esfahan	Max	1963	2010	1998	2010	2000	2002	1972	1972	2006	1967	1968	1995
(arid)	Min	1996	1972	1972	1972	1993	1972	2002	2002	2008	1996	2010	1982
Hamedan	Max	1967	2010	1961	2010	1964	1964	1986	2008	1968	1967	1982	2005
(semiarid)	Min	1986	1972	1964	1972	1961	1986	1967	1978	1964	1961	1964	1972
Jiroft	Max	2007	2001	1999	2001	2001	1998	1991	1991	1992	2007	1992	2001
(arid)	Min	1996	1992	1992	1992	1997	1995	1998	1998	2001	1996	2001	2009
Kerman	Max	1966	2010	2009	2010	1973	1970	1983	1974	1974	1969	1963	2010
(arid)	Min	1984	1972	1973	1972	1976	1992	1998	1962	2010	1984	2010	1983
Mashhad	Max	2010	2010	2006	2010	1980	2010	1982	1982	1976	1999	1992	2000
(semiarid)	Min	1981	1972	1972	1972	1991	1982	2010	2010	2006	1977	1983	1993
Moghan	Max	1989	2010	2010	2010	1989	1989	2003	2003	2003	1989	1988	1999
(semiarid)	Min	1994	1993	1993	2003	2003	1989	1989	1989	1996	1994	2004	1987
Qazvin	Max	1962	1966	1966	1966	1961	1966	1992	1972	1972	1976	1969	1963
(semiarid)	Min	1992	1974	1976	1974	1972	1992	1966	1994	2007	1998	1995	1992
Rasht	Max	1975	2010	2010	2010	1971	1971	1988	1988	1972	1975	1977	1995
(humid)	Min	1993	1972	1972	1969	1984	1988	1967	1971	2010	1988	1989	1974
Sanandaj	Max	1971	2010	2010	2010	1983	1970	1982	1982	1969	1971	1969	2001
(Mediterranean)	Min	1986	1992	1983	1992	1967	1992	1997	2010	1973	1986	2010	1986
Shahrekord	Max	2008	1978	1978	1978	2010	1964	1982	1980	2006	2008	1986	1973
(semiarid)	Min	1968	1992	2005	1992	1986	1992	1973	2010	2008	1968	1977	1984
Shiraz	Max	1981	1999	1999	2010	1973	2001	1969	1969	2004	1981	1992	2001
(semiarid)	Min	1992	1972	1968	1992	1996	1972	1966	2010	1966	2010	2008	1992
Tabriz	Max	1961	2010	2010	2010	2010	2010	1982	1982	1963	1961	1969	1999
(semiarid)	Min	1991	1972	1972	1972	1982	1982	2010	2010	1990	1992	1999	1969
Urmia	Max	2010	2010	2001	2010	2010	2001	1982	1982	1994	2007	1969	1962
(semiarid)	Min	1992	1982	1982	1982	1969	1982	2001	2001	2005	1984	2009	1982
Yazd	Max	1971	2010	2010	2010	1979	1970	1982	1972	1986	1971	1986	2010
(arid)	Min	1995	1972	1964	1972	1996	1986	2010	2010	2010	1995	1966	1971
Zabol	Max	1984	2006	2006	2010	1971	1971	1991	1982	2005	1984	1991	1983
(arid)	Min	1968	1972	1972	1972	1982	1991	1971	2010	2010	1968	1963	2003

If the annual extremum values of ET_0 and a particular meteorological variable coincide, they are highlighted by green color (Table 4).

As can be seen, the maximum of both ET_0 and *WS* in Ahvaz, Bushehr, Hamedan, Jiroft, Moghan, Rasht, Sanandaj, Shahrekord, Shiraz, Tabriz, Yazd, and Zabol happened in 1964, 1967, 1967, 2007, 1989, 1975, 1971, 2008, 1981, 1961, 1971, and 1984, respectively.

Similarly, the minimum of ET_0 and *WS* in Arak, Esfahan, Jiroft, Kerman, Moghan, Sanandaj, Shahrekord, Yazd, and Zabol occurred in 1966, 1996, 1996, 1984, 1994, 1986, 1968, 1995, and 1968, respectively.

Overall, in 83.3% (all arid, Mediterranean and humid regions and 66.7% of the semiarid regions) of the study sites, annual extremum values of ET_0 and *WS* occurred in the same year. These results imply that *WS* is the primary driver of the variations in ET_0 in Iran. In addition, the annual extremum values of ET_0 and *RH* coincided in 33.3% of the study sites, namely Ahvaz, Arak, Hamedan, Mashhad, Moghan, and Qazvin. Hence, *RH* is one of the main driving forces for variations of ET_0 in these study sites.

Figure 6 illustrates the frequency of maximum and minimum values of ET_0 and meteorological variables (namely *Tmean, Tmax, Tmin, Tmax-Tmin, es-ea, P, RH, RHmin, CD, WS, WD,* and *n*) in all the study sites for each year. Higher frequencies are observed in 1972, 1982, and 1992, indicating a large number of study sites reached their maximum and minimum values of ET_0 and meteorological variables in these years. This happened because 1972, 1982, and 1992 are El Niño-Southern Oscillation (ENSO) years [64–66]. The highest frequency during the 50-year period is observed in 2010. This year was considered as a dry year in Iran and most regions of Iran experienced drought alarms [50,52–57].

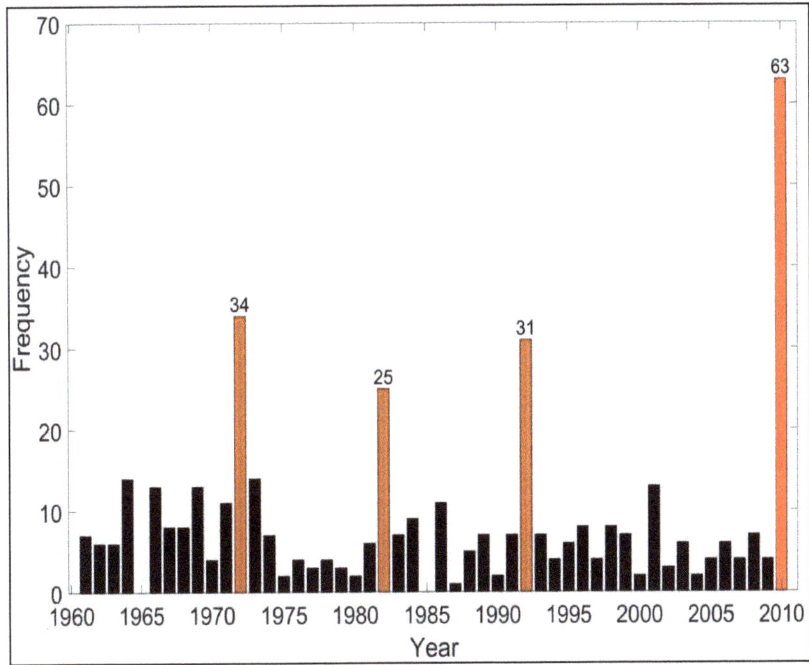

Figure 6. Frequency of maximum and minimum values of ET_0 and 12 meteorological variables in 18 study sites for each year.

4.6. Correlation between ET_o and Meteorological Variables

Figure 7 shows the correlation coefficient maps among the meteorological variables and reference evapotranspiration for the study sites in arid, Mediterranean, and humid regions. Similarly, Figure 8 indicates the correlation coefficient maps for the sites in semiarid region.

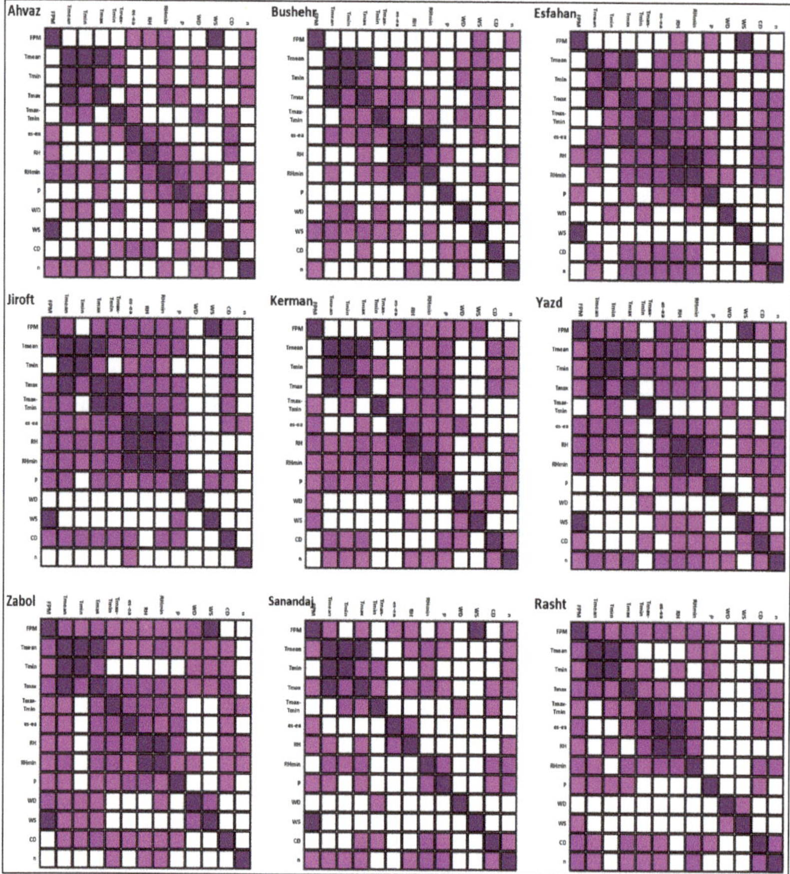

Figure 7. Spearman's Rho correlation coefficient (ρ) maps for weather stations located in arid, Mediterranean, and humid regions. Colored boxes show a significant correlation at the confidence level of 95%. Dark, medium, and light purples denote $0.8 \leq \rho \leq 1$, $0.5 \leq \rho < 0.8$, and $\rho < 0.5$, respectively.

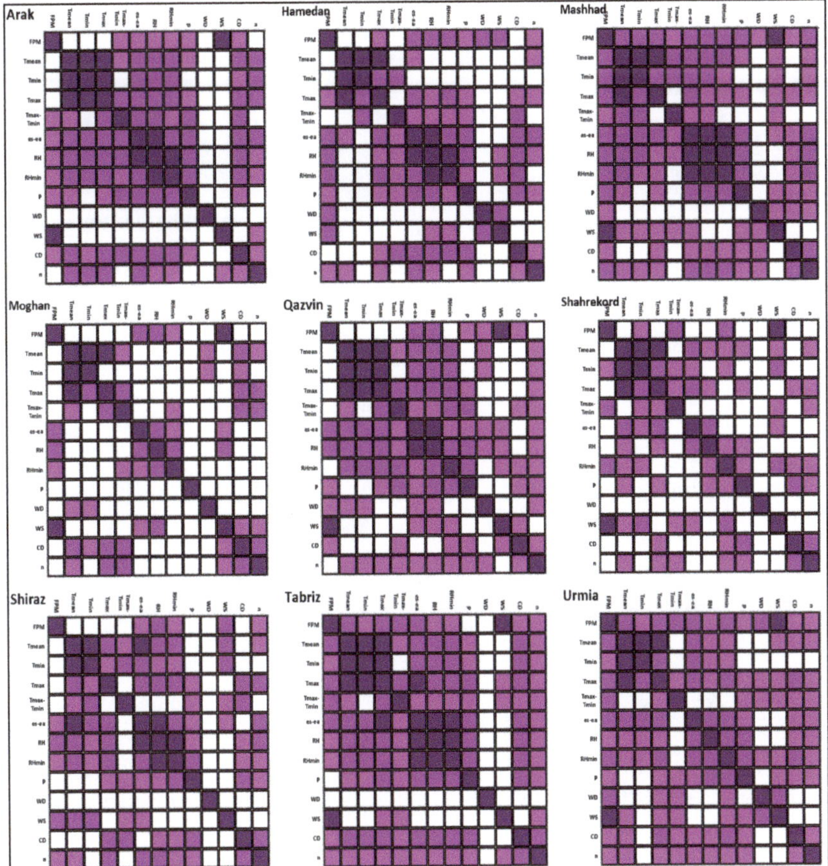

Figure 8. Spearman's Rho correlation coefficient (ρ) maps for weather stations located in semiarid region. Colored boxes show a significant correlation at the confidence level of 95%. Dark, medium, and light purples denote $0.8 \leq \rho \leq 1$, $0.5 \leq \rho < 0.8$, and $\rho < 0.5$, respectively.

As indicated, *WS* was the only meteorological variable that had high correlation ($\rho \geq 0.8$) with ET_0 in Ahvaz, Esfahan, Jiroft, Yazd, Zabol, Sanandaj, Arak, Mashhad, Moghan, Qazvin, Shahrekord, Tabriz, and Urmia. These results imply that *WS* was the driving force for the variations of ET_0 in most (72.2%) of the study sites in Iran, including 87.5% of arid regions and 77.8% of semiarid regions. This is in agreement with the results obtained by the TFPW-MK test (Table 3 and Figure 4).

WS and *WD* had the lowest correlation with the other meteorological variables. Hence, these variables may be controlled mainly by human activities such as desertification, land use change, and urban development [67–79]. The highest correlations were observed among the air temperature-related variables (i.e., *Tmean, Tmax, Tmin, Tmax-Tmin*) and humidity. In Mashhad and Urmia, the FPM ET_0 estimates had a good correlation with all of the meteorological variables. The upward trends of *Tmin* in all the study sites in Iran (except Shahrekord due to its high elevation) indicated a signal of climate change and signaled the need for the optimization of cropping pattern [80,81].

Table 5 shows the total number of meteorological variables in each study site with significant correlations with ET_0. It also lists the three meteorological variables that have the highest significant correlation with ET_0. As can be seen, ET_0 had the highest significant correlation with *WS* in all the sites except Rasht. ET_0 indicated the largest (the second largest) correlation with *Tmax* in Rasht (Mahshad, Tabriz, Urmia, and Zabol). According to Table 5, in 66.7% of the study sites, more than five meteorological variables had significant correlations with ET_0. This highlights the complexity of the forces driving variations of ET_0.

Table 5. The total number of meteorological variables in each site with significant correlations with ET_0, and the three meteorological variables with the highest significant correlation with ET_0.

Sites	Number of Significant Meteorological Variables	Three Highest-Ranked (from Left to Right) Significant Meteorological Variables
Ahvaz	5	WS, RHmin, RH
Arak	7	WS, es-ea, RH
Bushehr	2	WS, n
Esfahan	3	WS, P, RH
Hamedan	8	WS, es-ea, RH
Jiroft	9	WS, Tmax-Tmin, P
Kerman	7	WS, es-ea, RH
Mashhad	12	WS, Tmax, Tmin
Moghan	4	WS, RH, es-ea
Qazvin	6	WS, es-ea, RH
Rasht	11	Tmax, RHmin, es-ea
Sanandaj	7	WS, P, n
Shahrekord	4	WS, RHmin, Tmax-Tmin
Shiraz	5	WS, RH, RHmin
Tabriz	10	WS, Tmax, es-ea
Urmia	12	WS, Tmax, RHmin
Yazd	9	WS, RH, es-ea
Zabol	10	WS, Tmax, WD

In this study, the TFPW-MK and Spearman's Rho tests were used to identify significant (at the confidence levels 90%, 95% and 99%) trends of ET_0 and meteorological variables across Iran. These significant trends are indicative of climate change signals [1–24]. For instance, in the Mashhad site, significant upward (downward) trends were seen for ET_0, *Tmean*, *Tmin*, *Tmax*, and *es-ea* (*Tmax – Tmin*, *WD*, *RH* and *RHmin*), suggesting climate change alarms. These findings are consistent with those of other studies [82–87]. In the Shahrekord site, *Tmax – Tmin* (*Tmean*, *Tmin*, *RHmin*, and *CD*) showed significant increasing (decreasing) trends, which can be considered as climate change signals. These findings are also in agreement with those of [8,9,20–23]. In the Shiraz site, significant upward (downward) trends of ET_0, *Tmean*, *Tmin*, *Tmax*, and *es-ea* (*Tmax – Tmin*, *WS*, *RH* and *RHmin*) represented signals of climate change [88]. In the Urmia site, growing (reducing) trends of ET_0, *Tmax*, *WS*, *n*, *Tmean* and *Tmax* (*RHmin* and *CD*) signified climate change [89–92]. Finally, in the Zabol site, upward (downward) trends of ET_0, *Tmean*, *Tmin*, and *WS*, and *Tmax* (*RHmin*) may be considered as alarms of climate change [84,93].

The FPM ET_0 retrievals mainly depend on climate variables such as air temperature and humidity, wind speed, and solar radiation. This equation assumes a stomatal resistance of 70 s/m and an albedo of 0.23 for a standard 12 cm grass, and thus it ignores changes in stomatal resistance response to the elevated CO_2 and the surface albedo, ultimately causing uncertainty in the ET_0 estimates. Surface albedo varies with changes in vegetation and soil moisture [94].

Elevated CO_2 emission influences plant physiology by diminishing stomatal conductance [95–97]. Ignoring changes in the surface albedo and vegetation response to the elevated CO_2 emission lead to uncertainty in the FPM ET_0 estimates and results of this study. A similar uncertainty in the FPM ET_0 estimates was reported by other studies. For example, Milly and Dunne [98] showed that ignoring the stomatal resistance response to the increased CO_2 emission in the FPM equation led to discrepancy between the ET_0 estimates from climate models and the FPM equation. In a similar effort, Li et al. [99]

indicated that the variations in the FPM ET_0 estimates are only less than 10% for 40% changes in stomatal resistance. However, there are still high uncertainties in the amount of stomatal response to the raised CO_2 emission. For example, Domec et al. [100] reported that vegetation responses of pine to long term elevated CO_2 were manifested only when soil moisture was at a high level. Future studies should be directed toward evaluating the uncertainties in the FPM ET_0 estimates with respect to climate change and drought condition [101,102].

5. Conclusions

This study assessed the trends of meteorological variables (namely, mean, minimum, and maximum air temperature, difference between the maximum and minimum air temperature, mean and minimum relative humidity, wind speed and direction, vapor pressure deficit, rainfall, and sunshine hours) and reference evapotranspiration (ET_0) in the 18 study sites in different climate regions of Iran.

The effect of meteorological variables on ET_0 was also evaluated. Using the trend-free pre-whitening Mann–Kendall (TFPW-MK) and Spearman's Rho tests, wind speed (*WS*) was found to be the most important variable that controls the trend of ET_0 in most regions of Iran. Moreover, the increasing/decreasing trends of other meteorological variables (air temperature and humidity, rainfall, wind speed and direction, and sunshine hours) were indicative of climate change in many regions of Iran. The upward trend of minimum air temperature (*Tmin*) in all of the study sites in Iran (except Shahrekord due to its high elevation) indicated a signal of climate change and signaled the need for the optimization of cropping system. The significant increasing rate of *Tmin* may lead to a reduction in the growing degree day (GDD), ultimately decreasing the production of strategic and vital crops in future.

WS had significant upward and downward trends in 22.2% and 33.3% of the study sites, respectively. Therefore, significant variations in *WS* were observed in 55.5% of the investigated regions. The results also indicated that there were significant upward trends for mean air temperature (*Tmean*) and maximum air temperature (*Tmax*) in 77.8% and 66.7% of the study regions, respectively. In most of semiarid climates, *Tmin* and *Tmean* showed increasing, and minimum relative humidity (*RHmin*) and cloudy days (*CD*) indicated decreasing trends. This indicates that arid regions were affected more than semiarid areas by climate change. Precipitation (*P*), *WS*, and *CD* showed significant trends in 42.9%, 28.6%, and 28.6% of arid regions, respectively. However, *WS*, *Tmin*, *RHmin*, *P*, and *CD* had significant trends in 44.4%, 22.2%, 11.1%, 11.1%, and 11.1% of semiarid regions, respectively.

In 83.3% (all arid, Mediterranean and humid regions and 66.7% of semiarid regions) of the study sites, ET_0 and *WS* reached their maximum and/or minimum in the same year(s). Furthermore, ET_0 had the highest significant correlation with *WS* in all the sites except Rasht. This indicates that *WS* is the primary driver of the variations in ET_0 in most weather stations of Iran, especially in the years with extremum values of ET_0. However, in a number of weather stations, the significant correlation between ET_0 and other meteorological variables such as *Tmax*, *Tmax-Tmin*, *es-ea*, *P*, *RH*, *RHmin*, and *n* made it difficult to characterize driving forces for the trend of ET_0 and signals of climate change. The results of this study highlighted the complexity of forces that drive variations of ET_0.

Author Contributions: M.V. conceived and designed the study, performed the calculations, and prepared the original draft. S.M.B. supervised the study and revised the manuscript. M.A.G.S. and M.R.-S. supervised the study. V.P.S. reviewed the manuscript. All authors have read and agreed to the published version of the manuscript.

Funding: This research received no external funding.

References

1.   McNally, A.; Arsenault, K.; Kumar, S.; Shukla, S.; Peterson, P.; Wang, S.; Funk, C.; Peters-Lidard, C.D.; Verdin, J.P. A land data assimilation system for sub-Saharan Africa food and water security applications. *Sci. Data* **2017**, *4*, 170012. [CrossRef] [PubMed]

2. Li, M.; Xu, Y.; Fu, Q.; Singh, V.P.; Liu, D.; Li, T. Efficient irrigation water allocation and its impact on agricultural sustainability and water scarcity under uncertainty. *J. Hydrol.* **2020**, *586*, 124888. [CrossRef]

3. Abatzoglou, J.T.; Dobrowski, S.Z.; Parks, S.A.; Hegewisch, K.C. TerraClimate, a high-resolution global dataset of monthly climate and climatic water balance from 1958–2015. *Sci. Data* **2018**, *5*, 170191. [CrossRef] [PubMed]

4. McEvoy, D.J.; Huntington, J.L.; Mejia, J.F.; Hobbins, M.T. Improved seasonal drought forecasts using reference evapotranspiration anomalies. *Geophys. Res. Lett.* **2016**, *43*, 377–385. [CrossRef]

5. Yazdi, A.B.; Araghinejad, S.; Nejadhashemi, A.P.; Tabrizi, M.S. Optimal water allocation in irrigation networks based on real time climatic data. *Agric. Water Manag.* **2013**, *117*, 1–8.

6. Meza, F.J. Variability of reference evapotranspiration and water demands Association to ENSO in the Maipo River Basin. *Chile. Glob. Planet. Chang.* **2005**, *47*, 212–220. [CrossRef]

7. Hou, L.G.; Xiao, H.L.; Si, J.H.; Xiao, S.C.; Zhou, M.X.; Yang, Y.G. Evapotranspiration and crop coefficient of Populus euphratica Oliv forest during the growing season in the extreme arid region northwest China. *Agric. Water Manag.* **2010**, *97*, 351–356. [CrossRef]

8. Nouri, M.; Homaee, M.; Bannayan, M. Spatiotemporal reference evapotranspiration changes in humid and semi-arid regions of Iran: Past trends and future projections. *Theor. Appl. Climatol.* **2018**, *133*, 361–375. [CrossRef]

9. Nouri, M.; Homaee, M.; Bannayan, M. Quantitative trend, sensitivity and contribution analyses of reference evapotranspiration in some arid environments under climate change. *Water Resour. Manag.* **2017**, *31*, 2207–2224. [CrossRef]

10. Dadaser-Celik, F.; Cengiz, E.; Guzel, O. Trends in reference evapotranspiration in Turkey: 1975–2006. *Int. J. Climatol.* **2016**, *36*, 1733–1743. [CrossRef]

11. Song, Z.W.; Zhang, H.L.; Snyder, R.L.; Anderson, F.E.; Chen, F. Distribution and trends in reference evapotranspiration in the North China Plain. *J. Irrig. Drain. Eng.* **2010**, *136*, 240–247. [CrossRef]

12. Wang, W.; Peng, S.; Yang, T.; Shao, Q.; Xu, J.; Xing, W. Spatial and temporal characteristics of reference evapotranspiration trends in the Haihe River basin, China. *J. Hydrol. Eng.* **2010**, *16*, 239–252. [CrossRef]

13. Darshana; Pandey, A.; Pandey, R.P. Analysing trends in reference evapotranspiration and weather variables in the Tons River Basin in Central India. *Stoch. Environ. Res. Risk Assess.* **2013**, *27*, 1407–1421. [CrossRef]

14. Li, Z.X.; Qi, F.; Liu, W.; Wang, T.T.; Gao, Y.; Wang, Y.M.; Cheng, A.F.; Li, J.G.; Liu, L. Spatial and temporal trend of potential evapotranspiration and related driving forces in Southwestern China, during 1961–2009. *Quater. Int.* **2014**, *336*, 127–144.

15. Li, B.; Chen, F.; Guo, H. Regional complexity in trends of potential evapotranspiration and its driving factors in the Upper Mekong River Basin. *Quater. Int.* **2015**, *380-381*, 83–94. [CrossRef]

16. Gao, G.; Chen, D.; Ren, G.; Chen, Y.; Liao, Y. Spatial and temporal variations and controlling factors of potential evapotranspiration in China: 1956–2000. *J. Geograph. Sci.* **2006**, *16*, 3–12. [CrossRef]

17. Zhang, L.; Traore, S.; Cui, Y.; Luo, Y.; Zhu, G.; Liu, B.; Fipps, G.; Karthikeyan, R.; Singh, V. Assessment of spatiotemporal variability of reference evapotranspiration and controlling climate factors over decades in China using geospatial techniques. *Agric. Water Manag.* **2019**, *213*, 499–511. [CrossRef]

18. Liu, X.; Zhang, D. Trend analysis of reference evapotranspiration in Northwest China: The roles of changing wind speed and surface air temperature. *Hydrol. Process.* **2013**, *27*, 3941–3948. [CrossRef]

19. Liuzzo, L.; Viola, F.; Noto, L.V. Wind speed and temperature trends impacts on reference evapotranspiration in Southern Italy. *Theor. Appl. Climatol.* **2016**, *123*, 43–62. [CrossRef]

20. Khanmohammadi, N.; Rezaie, H.; Montaseri, M.; Behmanesh, J. The application of multiple linear regression method in reference evapotranspiration trend calculation. *Stoch. Environ. Res. Risk Assess.* **2018**, *32*, 661–673. [CrossRef]

21. Tabari, H.; Aeini, A.; Talaee, P.H.; Some'e, B.S. Spatial distribution and temporal variation of reference evapotranspiration in arid and semi-arid regions of Iran. *Hydrol. Process.* **2012**, *26*, 500–512. [CrossRef]

22. Tabari, H.; Abghari, H.; Hosseinzadeh Talaee, P. Temporal trends and spatial characteristics of drought and rainfall in arid and semiarid regions of Iran. *Hydrol. Process.* **2012**, *26*, 3351–3361. [CrossRef]

23. Kousari, M.R.; Ahani, H. An investigation on reference crop evapotranspiration trend from 1975 to 2005 in Iran. *Int. J. Climatol.* **2012**, *32*, 2387–2402. [CrossRef]

24. Shadmani, M.; Marofi, S.; Roknian, M. Trend analysis in reference evapotranspiration using Mann-Kendall and Spearman's Rho tests in arid regions of Iran. *Water Res. Manag.* **2012**, *26*, 211–224. [CrossRef]

25. Bonell, M.; Bruijnzeel, L.A. *Forests, Water and People in the Humid Tropics: Past, Present and Future Hydrological Research for Integrated Land and Water Management*; Cambridge University Press: London, UK, 2005.

26. Bormann, H. Sensitivity analysis of 18 different potential evapotranspiration models to observed climatic change at German climate stations. *Clim. Chang.* **2011**, *104*, 729–753. [CrossRef]

27. Kingston, D.G.; Todd, M.C.; Taylor, R.G.; Thompson, J.R.; Arnell, N.W. Uncertainty in the estimation of potential evapotranspiration under climate change. *Geophys. Res. Lett.* **2009**, *36*, L20403. [CrossRef]

28. Huo, Z.; Dai, X.; Feng, S.; Kang, S.; Huang, G. Effect of climate change on reference evapotranspiration and aridity index in arid region of China. *J. Hydrol.* **2013**, *492*, 24–34. [CrossRef]

29. Fan, J.; Wu, L.; Zhang, F.; Xiang, Y.; Zheng, J. Climate change effects on reference crop evapotranspiration across different climatic zones of China during 1956–2015. *J. Hydrol.* **2016**, *542*, 923–937. [CrossRef]

30. Zhang, D.; Liu, X.; Hong, H. Assessing the effect of climate change on reference evapotranspiration in China. *Stoch. Environ. Res. Risk Assess.* **2013**, *27*, 1871–1881. [CrossRef]

31. Dong, Q.; Wang, W.; Shao, Q.; Xing, W.; Ding, Y.; Fu, J. The response of reference evapotranspiration to climate change in Xinjiang, China: Historical changes, driving forces, and future projections. *Int. J. Climatol.* **2020**, *40*, 235–254. [CrossRef]

32. Lin, P.; He, Z.; Du, J.; Chen, L.; Zhu, X.; Li, J. Impacts of climate change on reference evapotranspiration in the Qilian Mountains of China: Historical trends and projected changes. *Int. J. Climatol.* **2018**, *38*, 2980–2993. [CrossRef]

33. Allen, R.G.; Pereira, L.S.; Raes, D.; Smith, M. *Crop Evapotranspiration-Guidelines for Computing Crop Water Requirements-FAO Irrigation and Drainage Paper 56*; FAO: Rome, Italy, 1998.

34. Bakhtiari, B.; Ghahraman, N.; Liaghat, A.M.; Hoogenboom, G. Evaluation of reference evapotranspiration models for a semiarid environment using lysimeter measurements. *J. Agric. Sci. Technol.* **2011**, *13*, 223–237.

35. Razzaghi, F.; Sepaskhah, A.R. Assessment of nine different equations for ETo estimation using lysimeter data in a semi-arid environment. *Arch. Agron. Soil Sci.* **2010**, *56*, 1–12. [CrossRef]

36. Razzaghi, F.; Sepaskhah, A.R. Calibration and validation of four common ET0 estimation equations by lysimeter data in a semi-arid environment. *Arch. Agron. Soil Sci.* **2012**, *58*, 303–319. [CrossRef]

37. Yarami, N.; Kamgar-Haghighi, A.A.; Sepaskhah, A.R.; Zand-Parsa, S. Determination of the potential evapotranspiration and crop coefficient for saffron using a water-balance lysimeter. *Arch. Agron. Soil Sci.* **2011**, *57*, 727–740. [CrossRef]

38. Mann, H.B. Non-parametric tests against trend. *Econometrica* **1945**, *13*, 163–171. [CrossRef]

39. Kendall, M.G. *Rank Correlation Methods*, 4th ed.; Charles Griffin: Oxford, UK, 1975.

40. Gilbert, R.O. *Statistical Methods for Environmental Pollution Monitoring*; Wiley: Hoboken, NJ, USA, 1987.

41. Sarr, M.A.; Gachon, P.; Seidou, O.; Bryant, C.R.; Ndione, J.A.; Comby, J. Inconsistent linear trends in Senegalese rainfall indices from 1950 to 2007. *Hydrol. Sci. J.* **2015**, *60*, 1538–1549. [CrossRef]

42. Park, K.I. *Fundamentals of Probability and Stochastic Processes with Applications to Communications*; Springer: Berlin, Germany, 2018.

43. Thiel, H. A rank-invariant method of linear and polynomial regression analysis (Part 3). In Proceedings of the Koninalijke Nederlandse Akademie van Weinenschatpen A, Amsterdam, The Netherlands, 25 February 1950; pp. 1397–1412.

44. Sen, P.K. Estimates of the regression coefficient based on Kendall's tau. *J. Am. Stat. Assoc.* **1968**, *63*, 1379–1389. [CrossRef]

45. Salas, J.D.; Delleur, J.W.; Yevjevich, V.; Lane, W.L. *Applied Modeling of Hydrologic Time Series*; Water Resources Publication: Littleton, CO, USA, 1980.

46. Coumou, D.; Robinson, A.; Rahmstorf, S. Glob. increase in record-breaking monthly-mean temperatures. *Clim. Chang.* **2013**, *118*, 771–782. [CrossRef]

47. Blain, G.C. Removing the influence of the serial correlation on the Mann-Kendall test. *Rev. Bras. Meteorol.* **2014**, *29*, 161–170. [CrossRef]

48. Henry, R.C.; Chang, Y.S.; Spiegelman, C.H. Locating nearby sources of air pollution by nonparametric regression of atmospheric concentrations on wind direction. *Atmos. Environ.* **2002**, *36*, 2237–2244. [CrossRef]

49. Sarmiento, C.; Valencia, C.; Akhavan-Tabatabaei, R. Copula autoregressive methodology for the simulation of wind speed and direction time series. *J. Wind Eng. Ind. Aerod.* **2018**, *174*, 188–199. [CrossRef]

50. Fitchett, J.M.; Grab, S.W.; Thompson, D.I.; Roshan, G. Spatio-temporal variation in phenological response of citrus to climate change in Iran: 1960–2010. *Agric. For. Meteorol.* **2014**, *198*, 285–293. [CrossRef]

51. Shirvani, A. Change point analysis of mean annual air temperature in Iran. *Atmos. Res.* **2015**, *160*, 91–98. [CrossRef]

52. Ashraf, B.; Yazdani, R.; Mousavi-Baygi, M.; Bannayan, M. Investigation of temporal and spatial climate variability and aridity of Iran. *Theor. Appl. Climatol.* **2014**, *118*, 35–46. [CrossRef]

53. Bazrafshan, J.; Nadi, M.; Ghorbani, K. Comparison of Empirical Copula-Based Joint Deficit Index (JDI) and Multivariate Standardized Precipitation Index (MSPI) for Drought Monitoring in Iran. *Water Resour. Manag.* **2015**, *29*, 2027–2044. [CrossRef]

54. Kehl, M. Quaternary climate change in Iran—the state of knowledge. *Erdkunde* **2009**, *63*, 1–17. [CrossRef]

55. Kousari, M.R.; Dastorani, M.T.; Niazi, Y.; Soheili, E.; Hayatzadeh, M.; Chezgi, J. Trend detection of drought in arid and semi-arid regions of Iran based on implementation of reconnaissance drought index (RDI) and application of non-parametrical statistical method. *Water Resour. Manag.* **2014**, *28*, 1857–1872. [CrossRef]

56. Tabari, H.; Nikbakht, J.; Talaee, P.H. Identification of trend in reference evapotranspiration series with serial dependence in Iran. *Water Res. Manag.* **2012**, *26*, 2219–2232. [CrossRef]

57. Rahimzadeh, F.; Nassaji Zavareh, M. Effects of adjustment for non-climatic discontinuities on determination of temperature trends and variability over Iran. *Int. J. Climatol.* **2014**, *34*, 2079–2096. [CrossRef]

58. Soltani, S.; Saboohi, R.; Yaghmaei, L. Rainfall and rainy days trend in Iran. *Climat. Chang.* **2012**, *110*, 187–213. [CrossRef]

59. Soltani, M.; Laux, P.; Kunstmann, H.; Stan, K.; Sohrabi, M.M.; Molanejad, M.; Sabziparvar, A.A.; SaadatAbadi, A.R.; Ranjbar, F.; Rousta, I.; et al. Assessment of climate variations in temperature and precipitation extreme events over Iran. *Theor. Appl. Climatol.* **2016**, *126*, 775–795. [CrossRef]

60. Hamzeh Nezhad, M.; Rabbani, M.; Torabi, T. The role of wind in human health in Islamic medicine and its effect in layout and structure of Iranian classic towns. *Naghsh Jahan* **2015**, *5*, 43–57.

61. Pirnia, M.K. *Stylistics of Iranian Architecture*; Sorush Danesh: Tehran, Iran, 2008.

62. Amiraslani, F.; Dragovich, D. Cross-sectoral and participatory approaches to combating desertification: The Iranian experience. *Natur. Resour. Forum* **2010**, *34*, 140–154. [CrossRef]

63. Amiraslani, F.; Dragovich, D. Combating desertification in Iran over the last 50 years: An overview of changing approaches. *J. Environ. Manag.* **2011**, *92*, 1–13. [CrossRef]

64. Roghani, R.; Soltani, S.; Bashari, H. Influence of southern oscillation on autumn rainfall in Iran (1951–2011). *Theor. Appl. Climatol.* **2016**, *124*, 411–423. [CrossRef]

65. Pourasghar, F.; Oliver, E.C.; Holbrook, N.J. Modulation of wet-season rainfall over Iran by the Madden–Julian Oscillation, Indian Ocean Dipole and El Niño–Southern Oscillation. *Int. J. Climatol.* **2019**, *39*, 4029–4040. [CrossRef]

66. Grove, R.; Adamson, G. *El Niño in World History*; Palgrave Macmillan: London, UK, 2018.

67. Azami, M.; Mirzaee, E.; Mohammadi, A. Recognition of urban unsustainability in Iran (case study: Sanandaj City). *Cities* **2015**, *49*, 159–168. [CrossRef]

68. Barzani, M.M.; Khairulmaini, O.S. Desertification risk mapping of the Zayandeh Rood Basin in Iran. *J. Earth Sys. Sci.* **2013**, *122*, 1269–1282. [CrossRef]

69. Emadodin, I.; Bork, H.R. Degradation of soils as a result of long-term human-induced transformation of the environment in Iran: An overview. *J. Land Use Sci.* **2012**, *7*, 203–219. [CrossRef]

70. Hashemimanesh, M.; Matinfar, H. Evaluation of desert management and rehabilitation by petroleum mulch base on temporal spectral analysis and field study (case study: Ahvaz, Iran). *Ecol. Eng.* **2012**, *46*, 68–74. [CrossRef]

71. Hosseini, S.M.; Sadrafshari, S.; Fayzolahpour, M. Desertification hazard zoning in Sistan Region, Iran. *J. Geograph. Sci.* **2012**, *22*, 885–894. [CrossRef]

72. Jafari, R.; Bakhshandehmehr, L. Quantitative mapping and assessment of environmentally sensitive areas to desertification in central Iran. *Land Degrad. Dev.* **2016**, *27*, 108–119. [CrossRef]

73. Khaledian, Y.; Kiani, F.; Ebrahimi, S. The effect of land use change on soil and water quality in northern Iran. *J. Mount. Sci.* **2012**, *9*, 798–816. [CrossRef]

74. Malmir, M.; Zarkesh, M.M.K.; Monavari, S.M.; Jozi, S.A.; Sharifi, E. Urban development change detection based on Multi-Temporal Satellite Images as a fast tracking approach—A case study of Ahwaz County, southwestern Iran. *Environ. Monitor. Assess.* **2015**, *187*, 1–10. [CrossRef]

75. Sadeghiravesh, M.H.; Ahmadi, H.; Zehtabian, G. Application of sensitivity analysis for assessment of de-desertification alternatives in the central Iran by using Triantaphyllou method. *Environ. Monit. Assess.* **2011**, *179*, 31–46. [CrossRef]

76. Sepehr, A.; Hassanli, A.M.; Ekhtesasi, M.R.; Jamali, J.B. Quantitative assessment of desertification in south of Iran using MEDALUS method. *Environ. Monit. Assess.* **2007**, *134*, 243–254. [CrossRef] [PubMed]

77. Tabari, H.; Somee, B.S.; Zadeh, M.R. Testing for long-term trends in climatic variables in Iran. *Atmos. Res.* **2011**, *100*, 132–140. [CrossRef]

78. Tabari, H.; Marofi, S.; Aeini, A.; Talaee, P.H.; Mohammadi, K. Trend analysis of reference evapotranspiration in the western half of Iran. *Agr. For. Meteorol.* **2011**, *151*, 128–136. [CrossRef]

79. Tabari, H.; Talaee, P.H.; Nadoushani, S.M.; Willems, P.; Marchetto, A. A survey of temperature and precipitation based aridity indices in Iran. *Quater. Int.* **2014**, *345*, 158–166. [CrossRef]

80. Eyshi Rezaie, E.; Bannayan, M. Rainfed wheat yields under climate change in northeastern Iran. *Meteorol. Appl.* **2012**, *19*, 346–354. [CrossRef]

81. Eyshi Rezaei, E.; Webber, H.; Gaiser, T.; Naab, J.; Ewert, F. Heat stress in cereals: Mechanisms and modelling. *Europ. J. Agron.* **2015**, *64*, 98–113. [CrossRef]

82. Banimahd, S.A.; Khalili, D.; Kamgar-Haghighi, A.A.; Zand-Parsa, S. In-depth investigation of precipitation-based climate change and cyclic variation in different climatic zones. *Theor. Appl. Climatol.* **2014**, *116*, 565–583. [CrossRef]

83. Farhangfar, S.; Bannayan, M.; Khazaei, H.R.; Baygi, M.M. Vulnerability assessment of wheat and maize production affected by drought and climate change. *Int. J. Disaster Risk Reduct.* **2015**, *13*, 37–51. [CrossRef]

84. Koocheki, A.; Nasiri, M.; Kamali, G.A.; Shahandeh, H. Potential impacts of climate change on agroclimatic indicators in Iran. *Arid Land Res. Manag.* **2006**, *20*, 245–259. [CrossRef]

85. Lashkari, A.; Alizadeh, A.; Rezaei, E.E.; Bannayan, M. Mitigation of climate change impacts on maize productivity in northeast of Iran: A simulation study. *Mitigat. Adapt. Strateg. Glob. Chang.* **2012**, *17*, 1–16. [CrossRef]

86. Moradi, R.; Koocheki, A.; Mahallati, M.N. Adaptation of maize to climate change impacts in Iran. *Mitigat. Adapt. Strateg. Glob. Chang.* **2014**, *19*, 1223–1238. [CrossRef]

87. Soltani, E.; Soltani, A. Climatic change of Khorasan, North-East of Iran, during 1950–2004. *Res. J. Environ. Sci.* **2008**, *2*, 316–322.

88. Abolverdi, J.; Ferdosifar, G.; Khalili, D.; Kamgar-Haghighi, A.A.; Haghighi, M.A. Recent trends in regional air temperature and precipitation and links to global climate change in the Maharlo watershed, Southwestern Iran. *Meteorol. Atmos. Physic.* **2014**, *126*, 177–192. [CrossRef]

89. Delju, A.H.; Ceylan, A.; Piguet, E.; Rebetez, M. Observed climate variability and change in Urmia Lake Basin, Iran. *Theor. Appl. Climatol.* **2013**, *111*, 285–296. [CrossRef]

90. Fathian, F.; Dehghan, Z.; Bazrkar, M.H.; Eslamian, S. Trends in hydrological and climatic variables affected by four variations of the Mann-Kendall approach in Urmia Lake basin, Iran. *Hydrol. Sci. J.* **2016**, *61*, 892–904. [CrossRef]

91. Fathian, F.; Morid, S.; Kahya, E. Identification of trends in hydrological and climatic variables in Urmia Lake basin, Iran. *Theore. Appl. Climatol.* **2014**, *119*, 443–464. [CrossRef]

92. Malekian, A.; Kazemzadeh, M. Spatio-temporal analysis of regional trends and shift changes of autocorrelated temperature series in Urmia Lake Basin. *Water Resour. Manag.* **2016**, *30*, 785–803. [CrossRef]

93. Valizadeh, J.; Ziaei, S.M.; Mazloumzadeh, S.M. Assessing climate change impacts on wheat production (a case study). *J. Saudi Soci. Agric. Sci.* **2014**, *13*, 107–115. [CrossRef]

94. Zhang, X.; Wang, W.C.; Fang, X.; Ye, Y.; Zheng, J. Agriculture development induced surface albedo changes and climatic implications across northeastern China. *Chin. Geogr. Sci.* **2012**, *22*, 264–277. [CrossRef]

95. Sun, S.; Chen, H.; Sun, G.; Ju, W.; Wang, G.; Li, X.; Yan, G.; Gao, C.; Huang, J.; Zhang, F.; et al. Attributing the changes in reference evapotranspiration in Southwestern China using a new separation method. *J. Hydrometeorol.* **2017**, *18*, 777–798. [CrossRef]

96. Field, C.B.; Jackson, R.B.; Mooney, H.A. Stomatal responses to increased CO_2: Implications from the plant to the global scale. *Plant Cell Environ.* **1995**, *18*, 1214–1225. [CrossRef]

97. Betts, R.A.; Boucher, O.; Collins, M.; Cox, P.M.; Falloon, P.D.; Gedney, N.; Hemming, D.L.; Huntingford, C.; Jones, C.D.; Sexton, D.M.; et al. Projected increase in continental runoff due to plant responses to increasing carbon dioxide. *Nature* **2007**, *448*, 1037–1041. [CrossRef]

98. Milly, P.C.; Dunne, K.A. Potential evapotranspiration and continental drying. *Nat. Clim. Chang.* **2016**, *6*, 946–949. [CrossRef]

99. Li, X.; Kang, S.; Niu, J.; Huo, Z.; Liu, J. Improving the representation of stomatal responses to CO_2 within the Penman-Monteith model to better estimate evapotranspiration responses to climate change. *J. Hydrol.* **2019**, *572*, 692–705. [CrossRef]

100. Domec, J.C.; Palmroth, S.; Ward, E.; Maier, C.A.; Thérézien, M.; Oren, R. Acclimation of leaf hydraulic conductance and stomatal conductance of Pinus taeda (loblolly pine) to long-term growth in elevated CO_2 (free-air CO_2 enrichment) and N-fertilization. *Plant Cell Environ.* **2009**, *32*, 1500–1512. [CrossRef] [PubMed]

101. Almazroui, M.; Saeed, F.; Saeed, S.; Islam, M.N.; Ismail, M.; Klutse, N.A.B.; Siddiqui, M.H. Projected Change in Temperature and Precipitation Over Africa from CMIP6. *Earth Syst. Environ.* **2020**, *4*, 455–475. [CrossRef]

102. Almazroui, M.; Saeed, S.; Saeed, F.; Islam, M.N.; Ismail, M. Projections of Precipitation and Temperature over the South Asian Countries in CMIP6. *Earth Syst. Environ.* **2020**, *4*, 297–320. [CrossRef]

Article

Increasing Neurons or Deepening Layers in Forecasting Maximum Temperature Time Series?

Trang Thi Kieu Tran [1], Taesam Lee [1,*] and Jong-Suk Kim [2,*]

[1]   Department of Civil Engineering, ERI, Gyeongsang National University, 501 Jinju-daero, Jinju 660-701,
     Korea; trangtran281292@gmail.com
[2]   State Key Laboratory of Water Resources and Hydropower Engineering Science, Wuhan University,
     Wuhan 430072, China
*   Correspondence: tae3lee@gnu.ac.kr (T.L.); jongsuk@whu.edu.cn (J.-S.K.)

Received: 8 September 2020; Accepted: 7 October 2020; Published: 9 October 2020

Abstract: Weather forecasting, especially that of extreme climatic events, has gained considerable attention among researchers due to their impacts on natural ecosystems and human life. The applicability of artificial neural networks (ANNs) in non-linear process forecasting has significantly contributed to hydro-climatology. The efficiency of neural network functions depends on the network structure and parameters. This study proposed a new approach to forecasting a one-day-ahead maximum temperature time series for South Korea to discuss the relationship between network specifications and performance by employing various scenarios for the number of parameters and hidden layers in the ANN model. Specifically, a different number of trainable parameters (i.e., the total number of weights and bias) and distinctive numbers of hidden layers were compared for system-performance effects. If the parameter sizes were too large, the root mean square error (RMSE) would be generally increased, and the model's ability was impaired. Besides, too many hidden layers would reduce the system prediction if the number of parameters was high. The number of parameters and hidden layers affected the performance of ANN models for time series forecasting competitively. The result showed that the five-hidden layer model with 49 parameters produced the smallest RMSE at most South Korean stations.

Keywords: artificial neural network; neurons; layers; temperature; South Korea; deep learning

---

1. Introduction

An artificial neural network (ANN) is a system for information processing inspired by biological neural networks. The key element of this network is the huge amount of highly interconnected processing nodes (neurons) that work together by a dynamic response to process the information. A neural network is useful for modeling the non-linear relation between the input and output of a system [1]. Compared to other machine learning methods such as autoregressive moving averages (ARMA), autoregressive integrated moving averages (ARIMA), and random forest (RF), the ANN model showed better performance in regression prediction problems [2–4]. According to Agrawal [5], the ANN model predicted rainfall events more accurately than the ARIMA model. In another work, ANNs have been applied to forecast monthly mean daily global solar radiation [6].

Furthermore, the ANN model has also been employed to forecast climatological and meteorological variables. Although it is known that the weather forecasting problem is challenging because of its chaotic and dynamic process, weather forecasting based on ANNs has been employing considerably in recent years due to the success of the ANN's ability. From some previous research, artificial neural networks have been shown as a promising method to forecast weather and time series data due to their capability of pattern recognition and generalization [7,8]. Smith et al. [9] developed an improved ANN to forecast the air temperature from 1 to 12 h ahead by increasing the number of samples in the training, adding additional seasonal variables, extending the duration of prior observations, and varying the number of hidden neurons in the network. Six hours of prior data were chosen as the inputs for the temperature prediction since a network with eight prior observations performed worse than the six hour network. Moreover, it is demonstrated that the models using one hidden layer with 40 neurons performed better than other models over repeated instantiations. In another study, the ANN models for the maximum as well as minimum temperature, and relative humidity forecasting were proposed by Sanjay Mathur [10] using time series analysis. The multilayer feedforward ANN model with a back-propagation algorithm was used to predict the weather conditions in the future, and it was found that the forecasting model could make a highly accurate prediction. The authors in [11] employed the ANN models to forecast air temperature, relative humidity, and soil temperature in India, showing that the ANN model was a robust tool to predict meteorological variables as it showed promising results with 91–96% accuracy for predictions of all cases. In this study, we also aimed to predict the air temperature one day ahead of past observations using the ANN model.

The effectiveness of a network-based approach depends on the architecture of the network and its parameters. All of these considerations are complex, and the configuration of a neural network structure depends on the problem [12]. If unsuitable network architecture and parameters are selected, the results may be undesirable. On the other hand, a proper design of network architecture and parameters can produce desirable results [13,14]. However, little investigation has been conducted on the effect of parameters and architecture on the model's performance. The selection of ANN architecture, consisting of input variables, the number of neurons in the hidden layer, and the number of hidden layers is a difficult task, so the structure of the network is usually determined by a trial and error approach and based on the experience of the modeler [5]. In another previous study, we compared one-hidden layer and multi-hidden layer ANN models in maximum temperature prediction at five stations in South Korea [15]. In addition, the genetic algorithm was applied to find the best architecture of models. It showed that the ANN with one hidden layer performed the most accurate forecasts. However, the effect of the number of hidden layers and neurons on the ANN's performance in the maximum temperature time series prediction is not sufficient. It may expect that the model performs worse when the number of parameters decreases. However, what happens if we further increase the number of tunable parameters? There are two competing effects. On the one hand, more parameters, which mean more neurons, become available, possibly allowing for better predictions. On the other hand, the higher the parameter number, the more overfitting the model is. Will the networks be robust if more trainable parameters than necessary are present? Is a one-hidden layer model always better than a multi-hidden layer model for maximum temperature forecasting in South Korea? Therefore, it is also apparently several problems related to the model proper architecture.

This paper proposed a new strategy that applied the ANNs using different learning parameters and hidden layers to empirically compare the prediction performance of daily maximum temperature time series. This study aimed to discuss the effect of parameters on the performance of ANN for temperature time series forecasting.

The rest of the paper is structured as follows. Section 2 describes the data and methodology used for the experiments. Section 3 describes the results, and the final section provides conclusions and directions for future work.

2. Data and Methods

In the current study, 55 weather stations that record maximum temperature in South Korea at the daily timescale were employed. Most stations have a data period of 40 years from 1976 to 2015, except for Andong station (1983–2015) and Chuncheon station (1988–2015).

Figure 1 presents the locations of the stations at which the data were recorded. The forecasting model for the maximum temperature was built based on the neural network. There were six neurons in the input layer, which corresponds to the number of previous days provided to the network for the prediction of the next maximum temperature value and one neuron in the output layer, respectively. The number of hidden layers and the number of hidden neurons are discussed. This study tested the performance of ANN models for one day ahead of the maximum temperature prediction using prior observations as inputs corresponding to three different cases of hidden layers; they were one, three, and five hidden layers, respectively. Besides, the following five levels of numbers of trainable parameters (i.e., the total number of its weights and biases) were selected for testing: 49, 113, 169, 353, and 1001. Combining the number of hidden layers and the number of parameters, Table 1 shows the model architectures. Besides, the configurations of 1-, 3- and 5-hidden layer ANN models with 49 learnable parameters are illustrated in Figure 2. It is noticed that the total number of trainable parameters was computed by summing the connections between layers and biases in every layer.

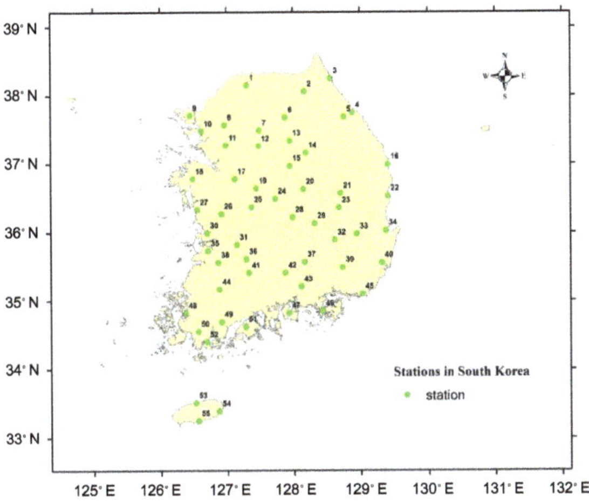

Figure 1. Map of the locations of the stations used in this study.

Table 1. Structure of the ANN models used for the study.

Number of Parameters	Number of Hidden Layers	Structure
49	1	6-6-1
	3	6-3-3-3-1
	5	6-5-1-1-1-1-1
113	1	6-14-1
	3	6-6-5-5-1
	5	6-4-4-4-4-4-1
169	1	6-21-1
	3	6-7-7-7-1
	5	6-6-6-4-4-6-1
353	1	6-44-1
	3	6-11-11-11-1
	5	6-8-8-8-8-8-1
1001	1	6-125-1
	3	6-20-20-20-1
	5	6-18-17-16-9-10-1

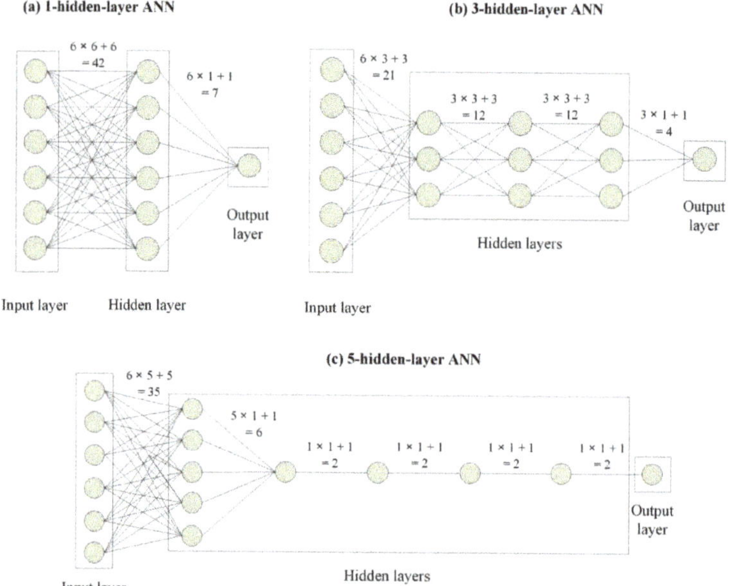

Figure 2. Schematic of the artificial neural network (ANN) with the total parameter number of 49 corresponding to (**a**) 1 hidden layer; (**b**) 3 hidden layers; and (**c**) 5 hidden layers.

To evaluate the effectiveness of each network, the root mean square error (RMSE) was used as a performance index. RMSE is calculated as

$$\text{RMSE} = \sqrt{\frac{1}{n}\sum_{i=1}^{n}(y_i - y'_i)^2} \tag{1}$$

where y_i is the observed data, y'_i is the predicted data, and n is the number of observations. The RMSE indicates a discrepancy between the observed and predicted data. The lower the RMSE, the more accurate the prediction is.

At each station, the data were subdivided into three parts: a training set consisting of 70% of the total data, a validation set using 20% of the total data, and a testing set containing 10% of the total data. The first two splits were used to train and examine the training performance, and the test set was used to evaluate the actual performance of the prediction. All models were trained and validated with the same training set and validation set, respectively. A popular technique to avoid the effect of overfitting the network on the training data is the early stopping method presented by Sarle [16]. Model overfitting is understood as the fitness of the model to the signal and noise that are usually present in the training sample. The possibility of overfitting depends on the size of the network, the number of training samples, and the data quality.

The maximum epoch for training was set at 1000. An epoch was described as one pass through all the data in the training set, and the weights of a network were updated after each epoch. The training was stopped when the error in the validation data reached its lowest value, or the training reached the maximum epoch, whichever came first. One iteration step during the ANN training usually works with a subset (call batch or mini-batch) of the available training data. The number of samples per batch (batch size) is a hyperparameter, defined as 100 in our case. Finally, the networks were evaluated based on the testing data. The ANN network was implemented using Keras [17], with TensorFlow [18] as the backend. According to Chen et al. [19], the design of neural networks has several aspects of concern. A model should have sufficient width and depth to capture the underlying pattern of the data. In contrast, a model should also be as simple as possible to avoid overfitting and high computational costs. However, the general trend of the number of parameters versus RMSE indeed provides some insights into selecting the proper structure of the model.

Data normalization is a preprocessing technique that transforms time series into a specified range. The quality of the data is guaranteed when normalized data are fed to a network. The MinMaxscaler technique was chosen for normalizing the data and making them in a range of [0, 1], which was defined as follows:

$$x' = \frac{x - x_{min}}{x_{max} - x_{min}} \tag{2}$$

where x' is the normalized data; x is the original data; and x_{max}, x_{min} are the maximum and minimum values of the data, respectively. The max and min used for standardization are calculated from the calibration period only. At the end of each algorithm, the outputs were denormalized into the original data to receive the final results.

An ANN consists of an input layer, one or more hidden layers of computation nodes, and an output layer. Each layer uses several neurons, and each neuron in a layer is connected to the neurons in the next layer with different weights, which change the value when it goes through that connection. The input layer receives the data one case at a time, and the signal will be transmitted through the hidden layers before arriving at the output layer, which is interpreted as the prediction or classification. The network weights are adjusted to minimize the output error based on the difference between the expected and target outputs. The error at the output layer propagates backward to the hidden layer until it reaches the input layer. The ANN models are used as an efficient tool to reveal a nonlinear relationship between the inputs and outputs [8]. Generally, the ANN model with three layers can be mathematically formulated as Lee et al. [20]:

$$y_k = f_2 \left[\sum_{j=1}^{m} W_{kj} f_1 \left(\sum_{i=1}^{n} W_{ji} x_i + b_j \right) + b_k \right] \tag{3}$$

where x_i is the input value to neuron i; y_k is the output at neuron k; f_1 and f_2 are the activation function for the hidden layer and output layer, respectively; n and m indicate the number of neurons in the input and hidden layers. W_{ji} is the weight between the input node i and hidden node j while W_{kj} is the weight between the hidden node j and output node k. b_j and b_k are the bias of the jth node in the hidden layer and the kth node in the output layer, respectively.

The weights in Equation (3) were adjusted to reduce the output error by calculating the difference between the predicted values and expected values using the back-propagation algorithm. This algorithm is executed in two specified stages, called forward and backward propagation. In the forward phase, the inputs were fed into the network and propagated to the hidden nodes at each layer until the generation of the output. In the backward phase, the difference between the true values and the estimated values or loss function was calculated by the network. The gradient of the loss function with respect to each weight can be computed and propagated backward to the hidden layer until it reaches the input layer [20].

In the current study, the 'tanh' or hyperbolic tangent activation function and an unthresholded linear function were used in the hidden layer and output layer, respectively. The range of the tanh function is from −1 to 1, and it is defined as follows:

$$\tanh(x) = \frac{e^x - e^{-x}}{e^x + e^{-x}} \tag{4}$$

It is noteworthy that there is no direct method well established for selecting the number of hidden nodes for an ANN model for a given problem. Thus, the common trial-and-error approach remains the most widely used method. Since ANN parameters are estimated by iterative procedures, which provide slightly different results each time they are run, we estimated each ANN 5 times and reported the mean and standard deviation errors in the figure.

The purpose of this study was not to find the best station-specific model, but to investigate the effects of hidden layers and trainable parameters on the performance of ANNs for maximum temperature modeling. We think that the sample size of 55 stations is large enough to infer some of the (average) properties of these factors to the ANN models.

3. Results

To empirically test the effect of the number of learnable parameters and hidden layers, we assessed and compared the model results obtained at 55 stations for five different parameters: 49, 113, 169, 353, and 1001, respectively. Moreover, we also tested the ANN models with different hidden layers (1, 3, and 5) having the same number of parameters at each station. Therefore, the mean and standard deviation were computed for the RMSE to analyze the impact of these factors on the ANN's performance. It is noted that all other modeling conditions, e.g., input data, activation function, number of epochs, and the batch size, were kept identical. After training and testing the datasets, the effects of the parameters and hidden layers of models were discussed.

3.1. Effect of the Number of Parameters

We first evaluated the performance of the ANN model by using the testing datasets. For each studied parameter, the prediction performance values were also presented as a rate of change in RMSE based on the original RMSE value obtained at 55 stations. This reference RMSE changes depending on the site used to study the impact of each parameter. Thus, as the results are depicted in Figures 3–5, the proposed ANN with 49 parameters consistently outperforms the other parameters at almost all stations in South Korea, since it produces the lowest change in error for different model configurations. Taking the single-hidden layer ANN as an example, the rate of change of the RMSE slightly increases with the extension of parameters for most sites (see Figure 3).

However, we can also observe that the increased parameter size of the ANN model made the error's change decrease lightly at Buan stations. Moreover, several sites have the lowest change values of RMSE at the parameter of 1001, such as Daegwallyeong, Gunsan, Hongcheon, and Tongyeong. These sites have an increasing trend of error when the number of tunable parameters in the network is raised from 49 to 169 (Hongcheon) or 353 (Daegwallyeong, Gunsan, and Tongyeong) before declining to the lowest point at 1001. Similarly, Figure 4 illustrates the general relationship between the total numbers of parameters versus the rate of change of RMSE on testing data in all 55 stations for three hidden layers. It can be noted from this figure that in the majority of stations, the rise of parameter numbers makes the performance of the model worse due to the increase in the change of RMSE. In contrast, few stations have the best results at the parameter of 133 (Pohang), 169 (Haenam), 353 (Buan and Yeongdeok), and 1001 (Daegwallyeong). Although the fluctuation of the RMSE's change rate for the three-hidden layer ANN, corresponding to various parameter sizes, varies from site to site, the ANN model with a structure of 49 trainable parameters still shows the best solution for predicting the maximum temperature one day ahead for most stations in South Korea.

In the case of five hidden layers, it can be observed from Figure 5 that the 49-parameter ANN model continues showing the smallest error in 52 out of the total 55 stations. In most cases, the increase in the number of parameters deteriorates the performance of the ANN model. However, it should be noticed that the model achieves the best result at the parameter of 353 in Buan and Gunsan stations while the smallest RMSE in Mokpo is obtained at the parameter of 119.

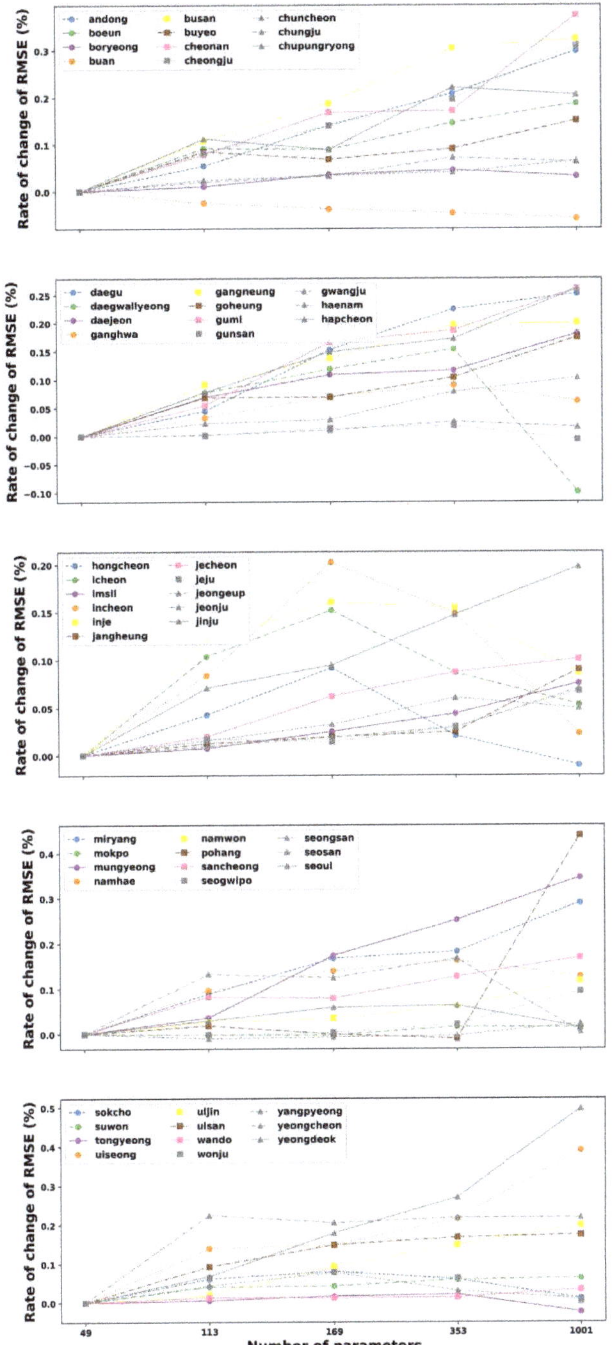

Figure 3. The rate of change of mean root mean square error (RMSE) of the 1-hidden layer ANN corresponding to 49, 113, 169, 353, and 1001 trainable parameters for the test data for all 55 stations in South Korea.

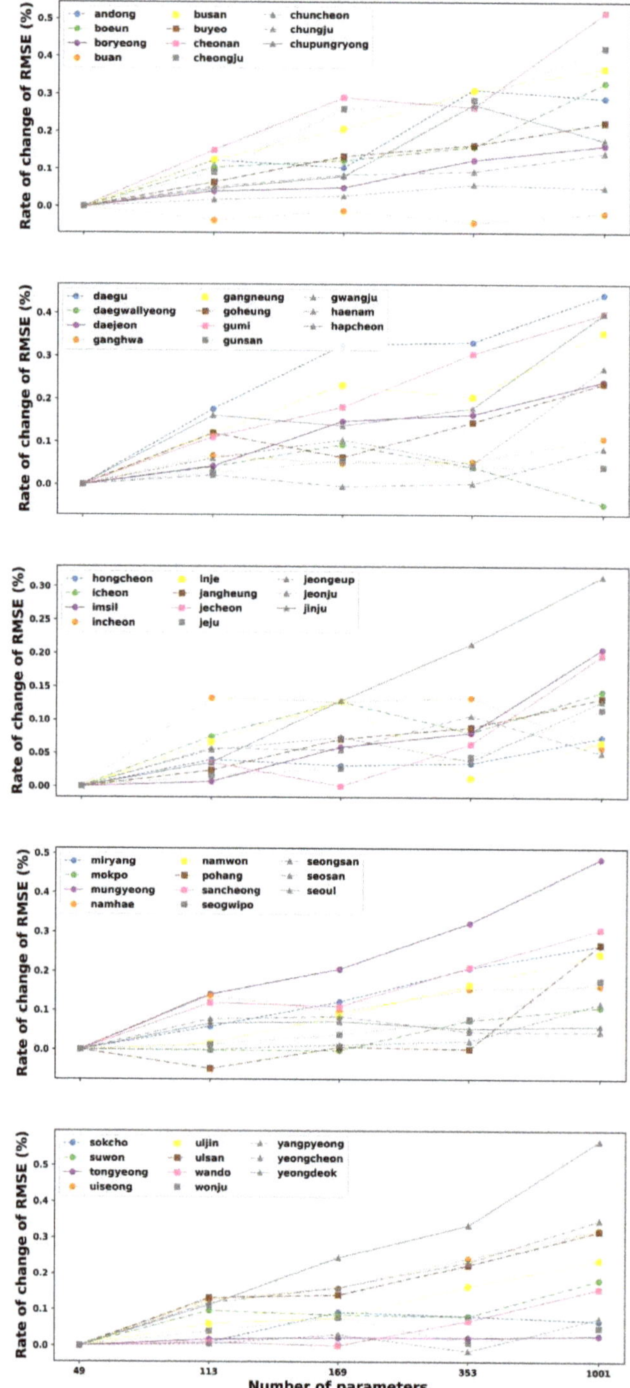

Figure 4. The rate of change of the mean RMSE of the 3-hidden layer ANN corresponding to 49, 113, 169, 353, and 1001 trainable parameters for the test data for all 55 stations in South Korea.

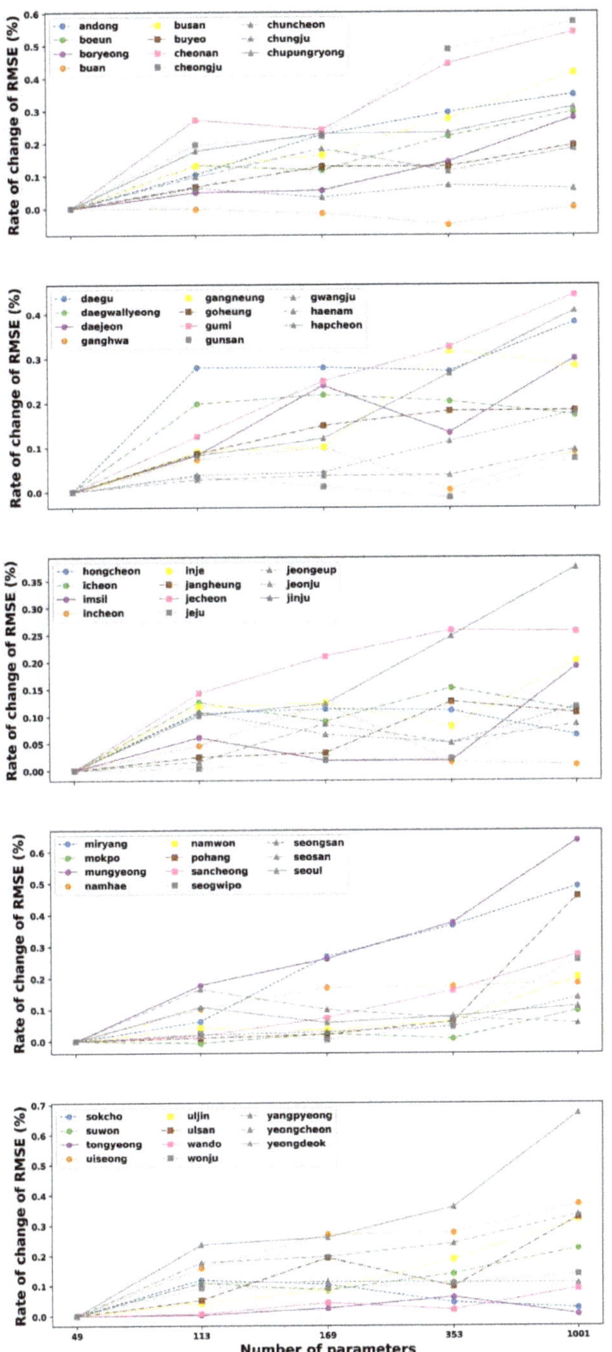

Figure 5. The rate of change of the mean RMSE of the 5-hidden layer ANN corresponding to 49, 113, 169, 353, and 1001 trainable parameters for the test data for all 55 stations in South Korea.

Figures S1–S4 (see Supplementary Materials) depict the RMSE of the ANN models with three different hidden layers that vary the number of parameters from 49 to 1001. The learnable parameter is an important parameter that may affect the ANN's performance for predicting the maximum temperature in the future. For a better assessment, the five-run average and standard deviation of the performances of each considered parameter and associated ANN architecture are depicted in those figures for each station separately. Generally, in most stations, there is a slight increasing trend of the RMSE when the number of parameters is increased. Furthermore, it can be seen from those figures that the ANN with parameters at 49 outperformed the parameters of other models, since it produced the lowest RMSE compared to others, except for Buan, Daegwallyeong (Figure S1), and Tongyeong stations (Figure S4), while the models having 1001 parameters yielded the worst results at around 50% of total stations, in comparison to other numbers of parameters. The difference in performance when the number of parameters increased from 49 to 1001 was marginal, but the value of 49 still leads to slightly better results. Thus, it can be noticed from these results that the testing RMSE of lower parameters is maybe better than higher parameters considering the low amount of neurons, and the models with 49 parameters are sufficient to forecast the maximum temperature variable. Nevertheless, there is an adaptive amount of parameters in the model in terms of hidden layers in some cases. For example, at Tongyeong stations (Figure S4), the model having one hidden layer showed the smallest RMSE at the parameters of 1001, while three and five-hidden layer models produced the best results with 49 parameters. In another case, 49 was the best number parameters for one and five hidden layers; meanwhile, three hidden layers presented the best performance at 353 parameters at Yangpyeong station (Figure S4). Besides, it is worth noting that at the same parameter number, the values of the RMSE for the one-, three-, and five-hidden layer models were comparable in most of the stations. However, the significant differences among the three configurations of the model can be observed at the 1001 parameters such as Boryeong, Cheongju, Daewallyeong (Figure S1), Mokpo (Figure S3), Wonju, and Yangpyeong stations (Figure S4) or the 353 parameters such as Ganghwa (Figure S1), Incheon, Jangheung (Figure S2), Seosan (Figure S3), and Yangpyeong stations (Figure S4). In addition, the RMSE shows little sensitivity to changes in the number of hidden layers in some stations, especially at high parameters due to the large fluctuation of standard deviation values. For example, in Tongyeong station (Figure S4), at the same parameter number of 353 or 1001, the standard deviation values of the RMSE of three- and five-hidden layer models are considerably larger than that of the one-hidden layer model. The trend is more evident for predicting the maximum temperature as the total number of parameters increased and occured in some stations, such as Buan, Daegwallyeong, Chuncheon (Figure S1), Wonju, or Yangpyeong stations (Figure S4). It can be suspected that the performance of the model may be significantly affected when the structure of the model becomes more complex. Based on the variation model performance in terms of hidden layers and parameters, it can be concluded that both the number of parameters and hidden layers were important to the model's performance, and the selection of parameters and hidden layers needs considerable attention because of the fluctuation in error.

3.2. Effect of the Number of Hidden Layers

Figures 6–10 show the spatial distribution of the ANN performances in the test period. Accordingly, a significant decrease in error is likely to move from the eastern to the western and southern part of South Korea with 49 learnable parameters (see Figure 6).

Similar spatial distributions of the changes in RMSEs also occur in Figures 7–10 when the number of parameters is increased to 113, 169, 353, and 1001. It can be concluded that the ANN models perform better in western and southern Korea (left panels). Moreover, the visualization of the differences in the RMSE between one hidden layer and three hidden layers (middle panels) as well as between one hidden layer and five hidden layers (bottom panels) at each station is also shown in Figures 6–10. It is noticed that with the same number parameters of 49, while one-hidden layer model presented better results than the three-hidden layer model at over 60% of total stations, the five-hidden layer model performed slightly greater than the one-hidden layer model at around 79% of stations where

RMSE differences greater than 0 were found (see Figure 6). However, with the increase in the number of parameters, the ANN models with one hidden layer produced better results than the three- and five-hidden layer models in almost all stations.

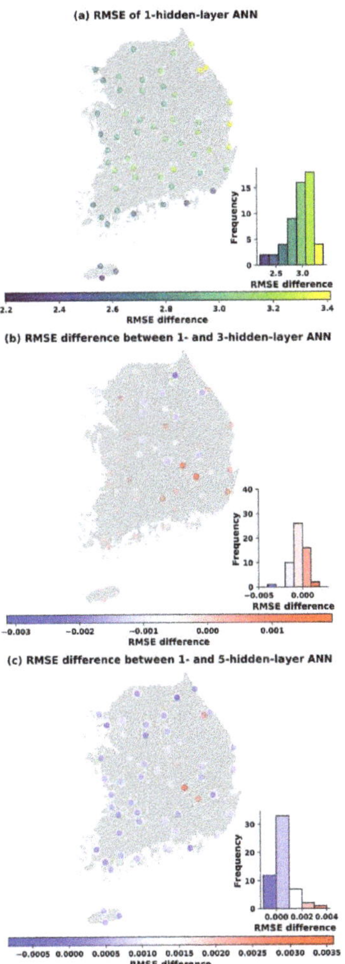

Figure 6. Panel (**a**) shows the RMSE of the test data of the 1-hidden layer model with 49 parameters; panel (**b**) shows the difference of the RMSE between the 1-hidden layer and the 3-hidden layer model (RMSE difference <0 indicates that the 1-hidden layer model performs better than the 3-hidden layer model and vice versa); and panel (**c**) shows the RMSE difference between the 1-hidden layer and the 5-hidden layer model (RMSE difference <0 indicates the that the 1-hidden layer model performs better, RMSE difference >0 shows the opposite).

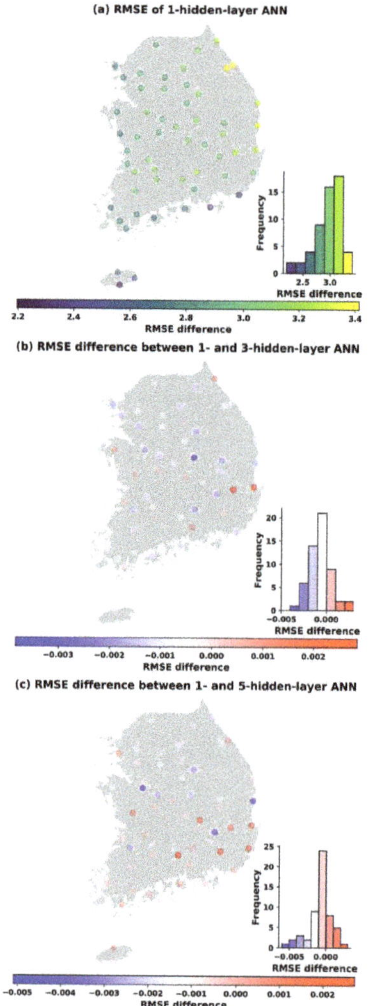

Figure 7. Panel (**a**) shows the RMSE of the test data of the 1-hidden layer model with 113 parameters; panel (**b**) shows the difference of the RMSE between the 1-hidden layer and the 3-hidden layer model (RMSE difference <0 indicates that the 1-hidden layer model performs better than the 3-hidden layer model and vice versa); and panel (**c**) shows the RMSE difference between the 1-hidden layer and the 5-hidden layer model (RMSE difference <0 indicates that the 1-hidden layer model performs better, RMSE difference >0 shows the opposite).

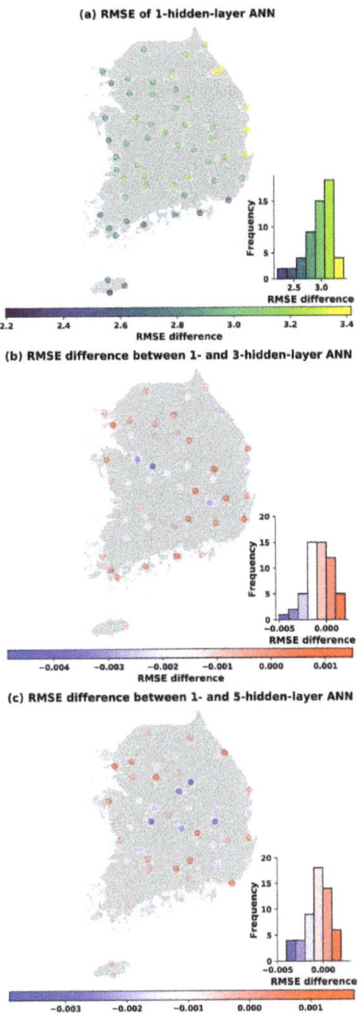

Figure 8. Panel (**a**) shows the RMSE of the test data of the 1-hidden layer model with 169 parameters; panel (**b**) shows the difference of the RMSE between the 1-hidden layer and the 3-hidden layer model (RMSE difference <0 indicates that the 1-hidden layer model performs better than the 3-hidden layer model and vice versa); and panel (**c**) shows the RMSE difference between the 1-hidden layer and the 5-hidden layer model (RMSE difference <0 indicates the that the 1-hidden layer model performs better, RMSE difference >0 shows the opposite).

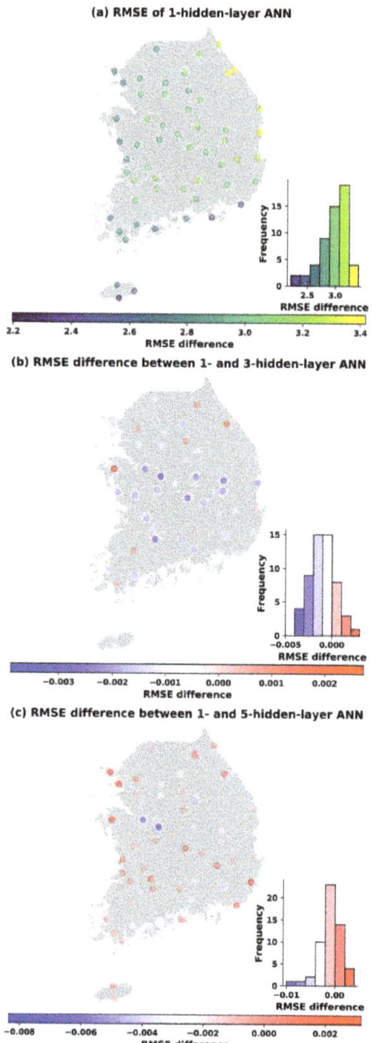

Figure 9. Panel (**a**) shows the RMSE of the test data of the 1-hidden layer model with 353 parameters; panel (**b**) shows the difference of the RMSE between the 1-hidden layer and the 3-hidden layer model (RMSE difference <0 indicates that the 1-hidden layer model performs better than the 3-hidden layer model and vice versa); and panel (**c**) shows the RMSE difference between the 1-hidden layer and the 5-hidden layer model (RMSE difference <0 indicates the that the 1-hidden layer model performs better, RMSE difference >0 shows the opposite).

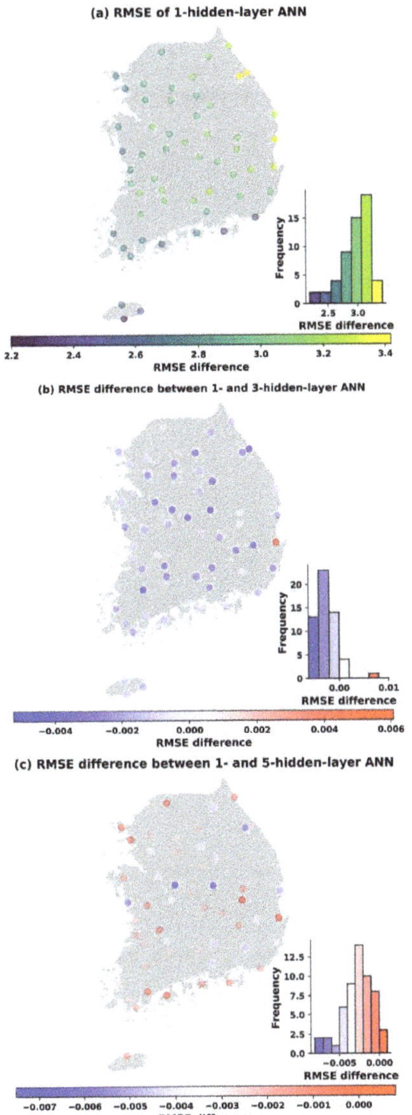

Figure 10. Panel (**a**) shows the RMSE of the test data of the 1-hidden layer model with 1001 parameters; panel (**b**) shows the difference of the RMSE between 1-hidden layer and the 3-hidden layer model (RMSE difference <0 indicates that the 1-hidden layer model performs better than the 3-hidden layer model and vice versa); and panel (**c**) shows the RMSE difference between the 1-hidden layer and the 5-hidden layer model (RMSE difference <0 indicates the that the 1-hidden layer model performs better, RMSE difference >0 shows the opposite).

For example, Figure 7 indicates that, based on the RMSE of the testing sets, the one-hidden layer model achieves a better performances at 42 and 41 stations, respectively, when the number of parameters is 113, compared to the three- and five-hidden layer models. Similarly, at the parameter number of 169, and from the histogram of RMSE differences (Figure 8), we can see that the one-hidden layer structure generally obtained a smaller error than three and five hidden layers for most sites.

However, it was noted that the multi-hidden layer models improved the performance of temperature prediction at some stations (17 stations with three hidden layers and 20 stations with five hidden layers) in comparison to one hidden layer. Moreover, the number of sites that obtained a smaller RMSE with multiple hidden layers decreased slightly when the number of parameters was increased to 353 (Figure 9). Out of 55 stations, there were 12 stations with three hidden layers having smaller errors than one hidden layer, and 18 stations with five hidden layers had better performance than one hidden layer. Finally, Figure 10 compares the ANN models in the test case in the name of the RMSE and hidden layer with 1001 trainable parameters. As the results showed, the one-hidden layer ANN generated a better result at almost every station than the three and five hidden layers. The three- and five-hidden layer ANNs had higher RMSE than one hidden layer in five and three stations out of 55 stations, respectively. The results were worth emphasizing that the deep ANN with various parameters trained for all stations generated a certain number of basins with lower performance than a single-hidden layer network, but the stations where this occurred were not always the same.

4. Summary and Conclusions

In this study, different hidden layer ANN models with various interior parameters were employed to forecast 1 day maximum temperature series over South Korea. This study aimed to explore the relationship between the size of the ANN model and its predictive capability, revealing that for future predictions of the time series of maximum temperature. In summary, the major findings of the present study are as follows.

Firstly, a deep neural network with more parameters does not perform better than a small neural network with fewer layers and neurons. The structural complexity of the ANN model can be unnecessary for unraveling the maximum temperature process. Even though the differences between these models are mostly small, it might be useful to some applications that require a small network with a lower computational cost because increasing the number of parameters may slow down the training process without substantially improving the efficiency of the network. Importantly, it can be observed that a simple network can perform better than a complex one, as also concluded in other comparisons. The authors in Lee et al. [20] showed that a large number of hidden neurons did not always lead to better performance. Similarly, in our previous study, the hybrid method of ANN and the genetic algorithm was applied to forecast multi-day ahead maximum temperature. The results demonstrated that the neural network with one hidden layer presented a better performance than the two and three hidden layers [14]. Nevertheless, too many or lesser amounts of parameters in the model can make the RMSE of the model increase, such as in the Buan station. This could be explained by an insufficient number of parameters causing difficulties in the learning data, whereas an excessive number of parameters might lead to unnecessary training time, and there is a possibility of over-fitting the training data set [21].

Secondly, although the performances of the models corresponding to different hidden layers are comparable when the number of parameters is the same, it is worth highlighting that five-hidden layer ANNs showed relatively better results compared to one and three hidden layers in the case of 49 parameters. However, when the number of parameters was large, the model with one hidden layer obtained the best solutions for forecasting problems in most stations.

Finally, the model's parameters and the degree of effectiveness of the hidden layers are relatively competitive in forecasting the maximum temperature time series due to variations in errors. Additionally, when the number of parameters is large, the significant difference of model outputs from various hidden layers can be achieved.

As future work, we are interested in investigating the effect of more parameters such as the learning rate or momentum on the system's performance. Moreover, conducting an intensive investigation on the effect of parameters on other deep learning approaches for weather forecasting, such as a recurrent neural network (RNN), long short-term memory (LSTM), and convolutional neural network (CNN) is a matter of interest for our future research. Besides, the sensitivity analysis or critical dependence of one parameter on others may be involved for further research.

Supplementary Materials: The following are available online at http://www.mdpi.com/2073-4433/11/10/1072/s1, Figure S1: RMSE of different parameters ANN corresponding to 1, 3, and 5 hidden layers for test data Andong, Boeun, Boryeong, Buan, Busan, Buyeo, Cheonan, Cheongju, Chuncheon, Chungju, Chupungryong, Daegu, Daewallyeong, Daejeon, and Ganghwa stations. The dot and the vertical lines denote the mean and the standard deviation of 5 repetitions, respectively, Figure S2: RMSE of different parameters ANN corresponding to 1, 3, and 5 hidden layers for test data at Gangneung, Goheung, Gumi, Gunsan, Gwangju, Haenam, Hapcheon, Hongcheon, Icheon, Imsil, Incheon, Inje, Jangheung, Jecheon, and Jeju stations. The dot and the vertical lines denote the mean and the standard deviation of 5 repetitions, respectively, Figure S3: RMSE of different parameters ANN corresponding to 1, 3, and 5 hidden layers for test data at Jeongeup, Jeonju, Jinju, Miryang, Mokpo, Mungyeong, Namhae, Namwon, Pohang, Sancheong, Seogwipo, Seongsan, Seosan, Seoul, and Sokcho stations. The dot and the vertical lines denote the mean and the standard deviation of 5 repetitions, respectively, Figure S4: RMSE of different parameters ANN corresponding to 1, 3, and 5 hidden layers for test data at Suwon, Tongyeong, Uiseong, Uljin, Ulsan, Wando, Wonju, Yangpyeong, Yeongchen, and Yeongdeok stations. The dot and the vertical lines denote the mean and the standard deviation of 5 repetitions, respectively.

Author Contributions: Conceptualization, formal analysis, writing—original draft, T.T.K.T.; conceptualization, resources, methodology, writing—original draft, T.L.; conceptualization, writing—original draft, writing—review and editing, J.-S.K. All authors have read and agreed to the published version of the manuscript.

Funding: This work was supported by the National Research Foundation of Korea (NRF) grant funded by the Korean Government (MEST) (2018R1A2B6001799).

Acknowledgments: The authors appreciate the journal editors and the reviewers for their useful and valuable comments for this study. We also thank the support of the State Key Laboratory of Water Resources and Hydropower Engineering Science, Wuhan University.

Conflicts of Interest: The authors declare no conflict of interest.

References

1. Valverde Ramírez, M.C.; De Campos Velho, H.F.; Ferreira, N.J. Artificial neural network technique for rainfall forecasting applied to the São Paulo region. *J. Hydrol.* **2005**, *301*, 146–162. [CrossRef]

2. Umakanth, N.; Satyanarayana, G.C.; Simon, B.; Rao, M.C.; Babu, N.R. Long-term analysis of thunderstorm-related parameters over Visakhapatnam and Machilipatnam, India. *Acta Geophys.* **2020**, *68*, 921–932. [CrossRef]

3. Verma, A.P.; Swastika Chakraborty, B. Now casting of Orographic Rain Rate Using ARIMA and ANN Model. In Proceedings of the IEEE 2020, Aranasi, India, 12–14 February 2020. [CrossRef]

4. Berbić, J.; Ocvirk, E.; Carević, D.; Lončar, G. Application of neural networks and support vector machine for significant wave height prediction. *Oceanologia* **2017**, *59*, 331–349. [CrossRef]

5. Agrawal, K. Modelling and prediction of rainfall using artificial neural network and ARIMA techniques. *J. Ind. Geophys. Union* **2006**, *10*, 141–151.

6. Ozoegwu, C.G. Artificial neural network forecast of monthly mean daily global solar radiation of selected locations based on time series and month number. *J. Clean. Prod.* **2019**, *216*, 1–13. [CrossRef]

7. Young, C.-C.; Liu, W.-C.; Hsieh, W.-L. Predicting the Water Level Fluctuation in an Alpine Lake Using Physically Based, Artificial Neural Network, and Time Series Forecasting Models. *Math. Probl. Eng.* **2015**, 1–11. [CrossRef]

8. Alotaibi, K.; Ghumman, A.R.; Haider, H.; Ghazaw, Y.M.; Shafiquzzaman, M. Future predictions of rainfall and temperature using GCM and ANN for arid regions: A case study for the Qassim region, Saudi Arabia. *Water (Switzerland)* **2018**, *10*, 1260. [CrossRef]

9. Smith, B.A.; Mcclendon, R.W.; Hoogenboom, G. Improving Air Temperature Prediction with Artificial Neural Networks. *Int. J. Comput. Inf. Eng.* **2007**, *1*, 3159.

10. Paras, S.; Kumar, A.; Chandra, M. A feature based neural network model for weather forecasting. *World Acad. Sci. Eng.* **2007**, *4*, 209–216.

11. Rajendra, P.; Murthy, K.V.N.; Subbarao, A.; Boadh, R. Use of ANN models in the prediction of meteorological data. *Model. Earth Syst. Environ.* **2019**, *5*, 1051–1058. [CrossRef]

12. Tsai, C.Y.; Lee, Y.H. The parameters effect on performance in ANN for hand gesture recognition system. *Expert Syst. Appl.* **2011**, *38*, 7980–7983. [CrossRef]

13. Hung, N.Q.; Babel, M.S.; Weesakul, S.; Tripathi, N.K. Hydrology and Earth System Sciences An artificial neural network model for rainfall forecasting in Bangkok, Thailand. *Hydrol. Earth Syst. Sci.* **2009**, *13*, 1413–1416. [CrossRef]

14. Tran, T.T.K.; Lee, T.; Shin, J.-Y.; Kim, J.-S.; Kamruzzaman, M. Deep Learning-Based Maximum Temperature Forecasting Assisted with Meta-Learning for Hyperparameter Optimization. *Atmosphere* **2020**, *11*, 487. [CrossRef]

15. Tran, T.T.K.; Lee, T. Is Deep Better in Extreme Temperature Forecasting? *J. Korean Soc. Hazard Mitig.* **2019**, *19*, 55–62. [CrossRef]

16. Sarle, W.S. Stopped Training and Other Remedies for Overfitting. Proc. 27th Symp. *Comput. Sci. Stat.* **1995**, *17*, 352–360.

17. Chollet, F. Home Keras Documentation. Available online: https://keras.io/ (accessed on 4 May 2020).

18. Abadi, M.; Agarwal, A.; Barham, P.; Brevdo, E.; Chen, Z.; Citro, C.; Corrado, G.S.; Davis, A.; Dean, J.; Devin, M.; et al. TensorFlow: Large-Scale Machine Learning on Heterogeneous Distributed Systems. *arXiv* **2016**, arXiv:1603.04467.

19. Chen, Y.; Tong, Z.; Zheng, Y.; Samuelson, H.; Norford, L. Transfer learning with deep neural networks for model predictive control of HVAC and natural ventilation in smart buildings. *J. Clean. Prod.* **2020**, *254*, 119866. [CrossRef]

20. Lee, J.; Kim, C.G.; Lee, J.E.; Kim, N.W.; Kim, H. Application of artificial neural networks to rainfall forecasting in the Geum River Basin, Korea. *Water (Switzerland)* **2018**, *10*, 1448. [CrossRef]

21. Alam, S.; Kaushik, S.C.; Garg, S.N. Assessment of diffuse solar energy under general sky condition using artificial neural network. *Appl. Energy* **2009**, *86*, 554–564. [CrossRef]

Publisher's Note: MDPI stays neutral with regard to jurisdictional claims in published maps and institutional affiliations.

Article

Integrated Flood Forecasting and Warning System against Flash Rainfall in the Small-Scaled Urban Stream

Jung Hwan Lee , **Gi Moon Yuk, Hyeon Tae Moon and Young-Il Moon ***

Urban Flood Research Institute, University of Seoul, Seoul 02504, Korea; jhlee88@uos.ac.kr (J.H.L.); gmoon@uos.ac.kr (G.M.Y.); hmoon@uos.ac.kr (H.T.M.)
* Correspondence: ymoon@uos.ac.kr

Received: 19 August 2020; Accepted: 7 September 2020; Published: 11 September 2020

Abstract: The flood forecasting and warning system enable an advanced warning of flash floods and inundation depths for disseminating alarms in urban areas. Therefore, in this study, we developed an integrated flood forecasting and warning system combined inland-river that systematized technology to quantify flood risk and flood forecasting in urban areas. LSTM was used to predict the stream depth in the short-term inundation prediction. Moreover, rainfall prediction by radar data, a rainfall-runoff model combined inland-river by coupled SWMM and HEC-RAS, automatic simplification module of drainage networks, automatic calibration module of SWMM parameter by Dynamically Dimensioned Search (DDS) algorithm, and 2-dimension inundation database were used in very short-term inundation prediction to warn and convey the flood-related data and information to communities. The proposed system presented better forecasting results compared to the Seoul integrated disaster prevention system. It can provide an accurate water level for 30 min to 90 min lead times in the short-term inundation prediction module. And the very short-term inundation prediction module can provide water level across a stream for 10 min to 60 min lead times using forecasting rainfall by radar as well as inundation risk areas. In conclusion, the proposed modules were expected to be useful to support inundation forecasting and warning systems.

Keywords: flood risk; urban flood forecasting and warning; inland-river combined flood system; LSTM

1. Introduction

Urban flooding is one of the most severe hazards in most cities worldwide. In particular, small-scale urban streams with very small watersheds and a flood travel time of 1 h to 3 h can cause severe damage to property and human life during heavy rainfall and floods. Rainstorms and floods have caused severe property damage and heavy casualties and hampered commercial activities, adversely affecting urban life [1,2]. Studies by the Intergovernmental Panel on Climate Change (IPCC) have shown that such extreme events will increase in frequency and intensity [3]. It has been observed that the frequency of localized heavy rain has been greatly increased and the scale of damage tends to be larger in the city of Seoul [4]. In 2020, 1 person died, 28 were marooned by the flood because of flash rainfall with an intensity of 57 mm/h at the location of Dorim stream, which is a small-scale stream in South Korea. At the time, the rainfall intensity did not show a large damage intensity scale, but a fatal accident occurred because preemptive actions such as flood forecasting, warning, and evacuation of dwellers were not performed well in that area [5]. Hence, it is necessary to mitigate flood damage through accurate and proactive urban forecasting of floods caused by small-scale urban streams. For the mitigation of flood disasters and damage in urban areas, it is necessary to estimate the damage that can be caused by rainfall runoff and reduce flood damage through structured and unstructured measures. The Seoul Metropolitan City in South Korea installed 529 river crisis management facilities such as automatic alarm facilities, text message boards, CCTVs, warning lights, emergency ladders, etc for 13 regional rivers. Moreover, civil servants and neighborhood patrols members have been designated officers in charge for evacuating people in the event of a heavy rainfall. Public servants operate the integrated disaster prevention system to acquire water level data for each municipality and stream in real time and broadcast flood warnings (depending on the water level), make advance announcements, provide riverside warnings, evacuate citizens from the riverside, issue a flood watch [6]. However, unlike flood forecasting in large inundation vulnerable areas, which is made using satellite and radar data from the Korea Meteorological Agency (KMA) and Han River Flood Control Office, flood forecasting for small- and medium-scale streams with a short flood travel time is still performed using real-time water level data, which makes the surrounding regions vulnerable to flood disasters in the event of heavy rainfall. In countries with well-established flood protection systems, studies have been actively conducted to systematically establish unstructured flood protection alternatives. Especially, the construction of real-time disaster warning systems has substantially raised concerns. Lin, et al. [7] develop an intelligent hydroinformatics integration platform (IHIP) to provide online forecasting of regional flood depths through the use of the latest hydroinformatics technologies such that actions can be efficiently taken to mitigate flood risks. And Li-Chiu, et al. [8] proposed a forecasting model which is composed of three steps: classification, point forecasting, and spatial expansion using classification module, SVM and spatial expansion module. While the use of radar measurements for urban hydrological applications has increased considerably in recent years [9–11], the accuracy of radar forecasts in small and medium cities remains limited because of uncertainties associated with rainfall prediction models [12]. Recently, artificial neural network (ANN) models have received considerable attention from scientists, and research has been conducted on the learning and prediction of hydrological time series data such as rainfall, discharge, and remote sensing data by various ANN models [13]. On the basis of a review of previous studies on the learning and prediction of hydrological time series data by various ANN models, we observed that recurrent neural networks (RNNs) show good performance when trained with long-term time series data [14,15]. Caihong et al. [16] employed ANN and Long Short-Term Memory (LSTM) for simulating the rainfall-runoff process on the basis of flood events and observed that the special units of the forget gate rendered the LSTM model better suited for the simulation and more intelligent than the ANN model. Furthermore, they found that the LSTM model was more stable than the ANN model and showed better simulation performance by considering different lead time modeling. Jinle et al. [17] used LSTM models for predicting precipitation on the basis of meteorological data after identifying the correlation between meteorological variables and precipitation. The prediction accuracy was improved by determining the relative importance of the input variables. Furthermore, Tianyu et al. [18] constructed T multivariate single-step LSTM networks using information on spatial and temporal dynamics of rainfall and early discharge to forecast flash floods in mountainous catchments. The most successful and frequently used RNN is the LSTM, which is trained with time series. The LSTM, a special type of RNN, is designed to overcome a drawback of the traditional RNN, namely, its incapability to learn long-term dependencies. Its development is regarded as a milestone in studies on time series problems in the field of machine learning. These previous researches, which have used hydrological time series data to train an RNN, achieved reliable prediction results by developing a neural network model, setting the hyperparameter, and compiling a large number of high-quality input data. This strategy can be applied for real-time flood forecasting for small and medium scale urban streams with a short flood travel time and for proactively devising flood protection measures.

Timely and accurate forecasts in flood-prone areas are essential prerequisites for the provision of reliable early warning systems. The objective of this paper is to develop the integrated inundation forecasting and warning system combined inland and river. It is an efficient real-time inundation forecasting and warning system, presenting the two type modules during flash rainfall periods, the short-term inundation prediction with a data-driven method, and the very short-term inundation prediction with a model-driven method.

In the short-term inundation prediction module, the water level data were used as input, and the LSTM model was used as the computational method to develop the point forecasting module to yield the inundation depth forecasts at control points. Forecasting the water level of the data-driven method used neural networks has shown the strength to calculate high accurate prediction results with long lead time if the high quality, consistency, and the enormous volumes of water level data were constructed. It can proactively convey powerful messages fastly and intuitively about warning of flash flooding to the public and civil servants to help decision-making on flood damage reduction.

Besides, in the very short-term inundation prediction module of the model-driven method, the results of the detailed predictions of urban flood-dynamics were presented. The great benefit of the physically-based model simulation is the ability to directly model inundations and impacts connected with the high spatially variability of rainfall forecast. However, the use of a model-driven method was always neglected due to some reasons such as high-resolution inundation modeling of large and complex cityscapes induce long computation times, and the requirement of special infrastructure and high-performance computers. Therefore, in this study, to address the highly dynamic nature of the rainfall-runoff model in urban areas, hydrodynamic simulations combined inland and river with automatic parameter estimation by DDS were implemented and to improve the long computation times, the automatic simplification module and 2-dimension inundation database are applied. Among all, hydrodynamic simulation combined inland and river is a critically essential process in urban areas because there are many drainage networks around the stream and it causes flash flood damage extremely especially in greatly urbanized.

This real-time urban flood forecasting and warning system based on the data-driven and model-driven method can be efficiently used to reduce flood risks endangering residents and properties by flash floods and sufficiently support synthetic decision-making for prevention. The combined system of the data-driven module using LSTM and the model-driven module using radar and rainfall-runoff models can enable an advanced warning of probable flash floods and regional inundation depths for disseminating alarms in flood-threatened areas with 10 min to 90 min lead time. These results show more preemptive and accurate flood forecasting and warning results with over 30 min lead time compared to current systems that rely on real-time data in the urban area, so it can be fully utilized for urban flood forecasting and citizen evacuation.

2. The Integrated Flood Forecasting and Warning System

A flowchart of the integrated flood forecasting and warning system is shown in Figure 1. The proposed system is composed of two steps: the short-term inundation prediction and the very short-term inundation prediction. Details of these two steps are described as follows.

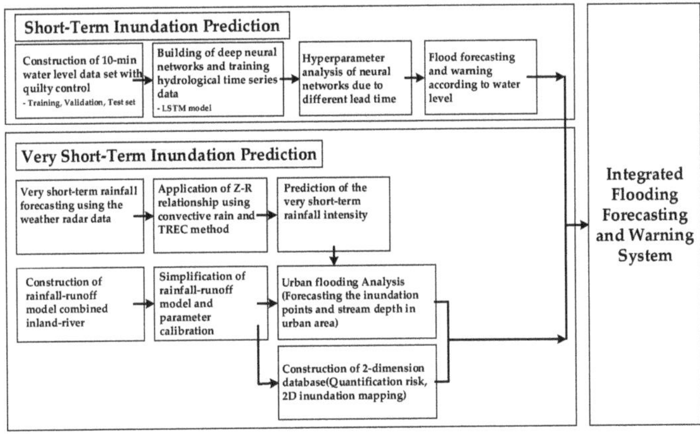

Figure 1. Schematic structure of the integrated flood forecasting and warning system for urban basin.

2.1. Short-Term Inundation Prediction

The Short-term inundation prediction module of the integrated flood forecasting and warning system is carried out by LSTM. The LSTM model proposed by Hochreiter and Schmidhuber [19] helps solve the gradient vanishing problem in an RNN by creating a path, through the introduction of self-loops, along which the gradient can flow for a long time. This implies that the weights can be changed at each time step by training the connection weights or by directly specifying them as hyperparameters, unlike the RNN (in which weights are fixed). The LSTM consists of multivariate single-step networks with information on spatial and temporal dynamics of rainfall and the water level as inputs. This module has predicted the urban stream depth by training the time series data of upstream depths. In order to determine the optimal prediction result, various training cases of LSTM were set up to forecast stream depth of target point according to training conditions of water level. Furthermore, the effects of the various inputs on urban flood forecasts for different lead times were analyzed through a parameter analysis. The lead times for each training group were set up in the range 30 min to 90 min for forecasting the water level at the target point.

The programming language of Python 3.7 is chosen with other libraries, such as Scikit-learn, Keras, Pandas, Numpy, and so on and the LSTM model is developed with the Deep-Learning framework TensorFlow 2.0 using jupyter notebook. The hyperparameter for each network model was calibrated automatically by a random search method of Keras tuner after specifying an appropriate range for the variables, instead of a grid search method, which is time-consuming. Grid search and manual search are the most widely used strategies for hyperparameter optimization. But, empirically and theoretically that randomly chosen trials are more efficient than grid search [20]. Granting random search the same computational budget, random search finds better hyperparameters by effectively searching a larger, less promising configuration space. We define the LSTM with 10 neurons in the first hidden layer and 5 neurons in the second hidden layer. The optimal hyperparameters were calculated for the validation dataset and the forecast result was analyzed for the test set. The mean square error (MSE) was used as a loss function, Rectified Linear Unit (ReLU) was used as a activation function and the Adam technique was chosen for the optimization algorithm with 0.001 [21]. Furthermore, He initialization was chosen to avoid the overfitting problem [22], and L2 regularization with 0.01, early stopping was also employed.

2.2. Very Short-Term Inundation Prediction

The critical technology of the very short-term inundation prediction module is divided into five parts: (A) forecasting rainfall using radar data; (B) constructing rainfall-runoff model of drainage networks combined inland-river; (C) simplification of the rainfall-runoff model using automatic simplification module; (D) parameter calibration of Storm Water Management Model (SWMM) using DDS and (E) setting up the 2-dimension inundation database. In the first step, the short-term rainfall forecasting using the radar data was implemented. It can be estimated for 10 min to 120 min before rainfall prediction with 10 min lead time using the Tracking Radar Echoes by Correlation (TREC) technique. In the second step, the rainfall-runoff model of drainage networks combined inland-river was built using coupled SWMM and the Hydrologic Engineering Center—River Analysis System (HEC-RAS). In the third step, the rainfall-runoff model, which is made the whole drainage networks in Dorim, was simplified by automatic simplification method. In the final step, the parameters of the SWMM model were calibrated automatically using a DDS to improve parameter uncertainty. The 2-dimension inundation database was built in advance by the rainfall-runoff model made of whole drainage networks in an urban area based on different rainfall and durations scenarios and it has been chosen and provided according to prediction rainfall by radar.

2.2.1. Rainfall Forecasting by Radar

The very short-term rainfall forecasting module of integrated flood forecasting and warning system was implemented using the radar data of Korea Meteorological Administration (KMA). KMA is operating 11 radars for the purpose of monitoring and forecasting dangerous weather conditions, flooding, and monitoring the military operation area. In this study, the Gwanak mountain meteorological radar was selected. The Gwanaksan meteorological radar has been installed in the center of the metropolitan area, playing a very important role in monitoring and forecasting dangerous weather. In addition, it has been upgraded to the dual-polarization radar from the single-polarization radar to improve rainfall prediction performance in urban areas in 2016. The rainfall prediction module forecasts rainfall up to 120 min at 10 min intervals using radar data. To this end, a Constant Altitude Plan Position Indicator (CAPPI) calculation program was developed, and the optimal movement patterns of intensive rainfall events using radar data are analyzed, and quantitative rainfall estimation is performed through the Z-R relational equation. The Tropical Z-R relationship ($Z = 334R^{1.19}$) has applied as an optimal radar Z-R relation, which is confirmed that the accuracy is improved over 20 mm/h heavy rainfall [23]. In order to predict the direction of rainfall in weather radar data, TREC is used. TREC is the first kind of radar-based nowcasting method, proposed by Rinehart and Garvey [24]. It is an image processing algorithm that calculates correlation coefficients between successive images of radar echoes and uses the maximum values to obtain the motion vectors of different regions. During calculations, if the regions for determining the correlation coefficient is set too large, the average moving direction and moving speed of the entire rainfall are obtained. On the contrary, if the regions are set too small, it is not easy to obtain meaningful results. In addition, the spatial position (calculation radius) between the two regions reflects the maximum distance of rainfall, so it must be determined within a physically meaningful range. Therefore, in this study, the size of the region was set to 21 km, and the size of the calculated radius was set to 7 km, reflecting the values suggested in the study by Kim and Kim [25]. Through this process, the linear motion of the rainfall during the targeted prediction time was predicted using the calculated motion vector field and reflectivity data. The function of the Z-R relationship algorithm is written as:

$$dBZ = 10logZ \tag{1}$$

$$Z = \sum D^6 \frac{mm^6}{m^3} \tag{2}$$

$$Z = aR^b \frac{mm^6}{m^3} \tag{3}$$

where Z is the factor of radar reflection (mm^6/m^3); R, rainfall intensity (mm/h); a and b, experienced constants.

2.2.2. Construction of the Rainfall-Runoff Model

Reliable drainage network modeling is an essential component in urban flood forecasting and risk assessment. Urban inundation has occurred attribute to a combination of river flooding and poor drainage of inland. In fact, in the case of urban basins, there are many drainage networks around the stream, which causes flooding frequently. However, most of the prospective studies on flooding have not been able to organically link urban inundation with river flooding and inland flooding. To improve this condition, there is a need for a combination of river channel geometry and sewage channel system. Therefore, we constructed the rainfall-runoff model of drainage networks combined inland-river. The hydraulic flood elevation routing model of HEC-RAS combined into the sewage network system of SWMM to consider the correlation between the water level of the stream and inland drainage facilities in the urban area. This rainfall-runoff module base on the Dynamic wave method of flood routing in SWMM is then coupled with HEC-RAS, leading to an integrated drainage network model. Figure 2 shows the process of constructing the rainfall-runoff model.

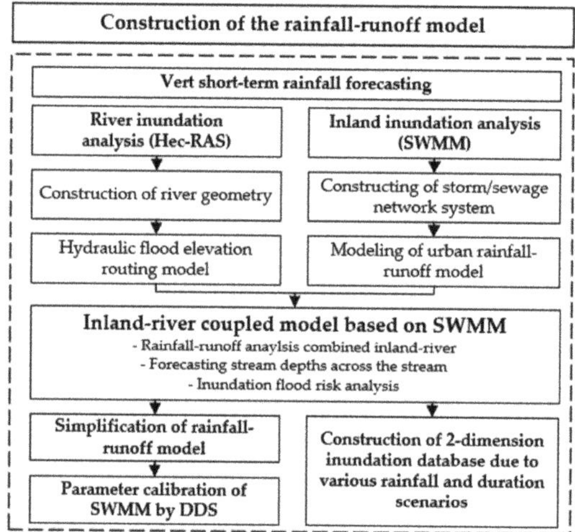

Figure 2. Process of constructing the rainfall-runoff model.

Not only that, Drainage network simplification by automatic simplification technology has been implemented to apply this module to real-time urban flooding. The rainfall-runoff model made of whole drainage networks in the major urban area such as Seoul is vast and complex, making it unsuitable for real-time urban flood forecasting. Therefore, the rainfall-runoff model should be simplified. The results of the simplified model showed that the integrated rainfall-runoff module can model and interpret flows with greatly decreased computational time at various channel sections and nodes in common with a not simplified model. And the simplified model has effectively given catchment responses for peak flow and volume of runoff which is considered as one of the essential components of urban drainage planning to mitigate the risk of flood. Figure 3 and Table 1 show the process of drainage network simplification and the simplified rainfall-runoff model of SWMM.

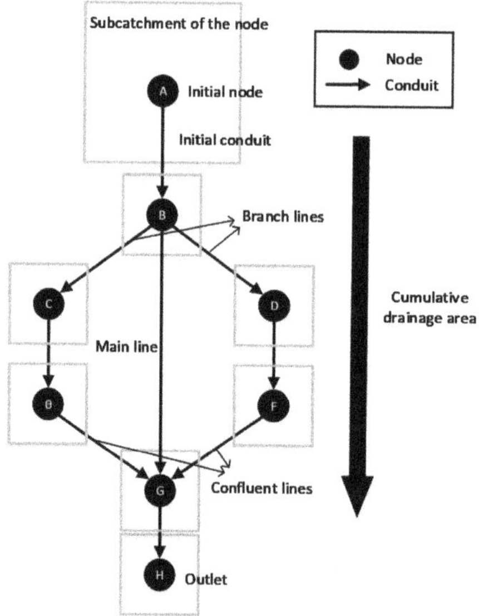

Figure 3. Drainage networks simplification structure.

Table 1. Process of drainage networks simplification.

Step	Detail Process
1st step Checking of the initial condition	Searching the initial conduit and node Calculating the cross-sectional area of flow Checking the branch conduits and nodes Checking the outlet
2nd step Calculating of the drainage area	Calculating the cumulative drainage area of all nodes from upstream point
3rd step Calculating of the branch line and mainline	User can define the cumulative drainage area to distinguish branch line and main line
4th step Calculating of the parameter	Calculating the parameters of nodes and conduits to be deleted in simplification process
5th step Building of the drainage network	Building the simplified drainage network(.inp)

A global optimization algorithm, DDS, is introduced for automatic calibration of watershed simulation models. SWMM can simulate the rainfall-runoff characteristics of an urban area using various parameters such as the diversity of land use, topography and the high spatial variability of rainfalls, but this model contain effective physical and conceptual model parameters which are either difficult or impossible to directly measure. Therefore, to apply these models, model parameters have to be adjusted to fit the observed runoff results. DDS selected as the optimization algorithm of the integrated rainfall-runoff module is designed for calibration problems with many parameters, requires no algorithm parameter tuning, and automatically scales the search to find good solutions within the maximum number of user-specified function evaluations [26]. Table 2 shows the DDS algorithm.

Table 2. Dynamically dimensioned search (DDS) algorithm (Tolson, B.A. and Shoemaker, C.A., 2007).

Step	Detail Process
1st step	Define DDS input: Neighborhood perturbation size parameter, γ (0.2 is default) Maximum # of function evaluation, m Vectors of lower, X^{min}, and upper, X^{max}, bounds for all D decision variables Initial solution, $X^0 = [X_1, \cdots, X_D]$
2nd step	Set counter to 1, $i-1$, and evaluate objective function F at initial solution, $F(X^0)$: $F_{best} = F(X^0)$, and $X^{best} = X^0$
3rd step	Randomly select J of the D decision variables for inclusion in neighborhood, {N}: Calculate probability each decision variable is included in {N} as a function of the current iteration count: $P(i) = 1 - \ln(i)/\ln(m)$ FOR d = 1, $\cdots$ D decision variables, add d to {N} with probability P IF {N} empty, select one random d for {N}
4th step	For j = 1, $\cdots$ J decision variables in {N}, perturb X_j^{best} using a standard normal random variable, $N(0,1)$, reflecting at decision variable bounds if necessary: $X_j^{new} = X_j^{best} + \sigma_j N(0,1)$, where $\sigma_j = Y(X_j^{max} - X_j^{min})$ IF $X_j^{new} < X_j^{min}$, reflect perturbation : $X_j^{new} = X_j^{min} + (X_j^{min} - X_j^{new})$ IF $X_j^{new} > X_j^{max}$, set $X_j^{new} = X_j^{min}$ IF $X_j^{new} > X_j^{max}$, reflect perturbation : $X_j^{new} = X_j^{max} - (X_j^{new} - X_j^{max})$ IF $X_j^{new} < X_j^{min}$, set $X_j^{new} = X_j^{max}$
5th step	Evaluate $F(X^{new})$ and update current best solution if necessary : IF $F(X_j^{new}) \leq F_{best}$, update new best solution : $F_{best} = F(X^{new})$ and $X^{best} = X^{new}$
6th step	Update iteration count, i = i+1, and check stopping criterion: IF i = m, STOP, print output (e.g., F_{best} & X^{best}) ELSE go to STEP 3

2D inundation numerical model is a widely used tool for flood inundation mapping. However, even high-performance computing equipment has spent substantial computing time running a 2-dimensional numerical model, so it is not appropriate the real-time inundation forecasting and warning. We have built an inundation database for various rainfall scenarios using Two-dimensional Unsteady FLOW (TUFLOW) engine and the inundation potential database was planned to be applied in emergency management according to similar rainfall patterns. The flash floods that result in serious damages frequently occur about 1 h to 24 h duration; thus, various probable rainfall and the design rainfall pattern of 1 h to 24 h duration were applied for numerical simulations. Furthermore, Huff's method [27] was employed to estimate the temporal distribution of the design rainfall for a watershed. Through this module, the emergency managers can easily compare the current rainfall conditions with the design rainfalls to estimate the flood extents and depths by using the inundation potential database during floods.

3. Study Area and Data Processing

3.1. Study Area and Runoff Model

The Seoul Metropolitan City manages a total of 34 inundation-vulnerable areas. Among them, Dorim basin, which was chosen as the target watershed, is the one of the inundation-vulnerable districts and has small-scale stream. Dorim basin has an area of 42.5 km^2 and the Dorim stream has a length of 14.51 km (Figure 4). Upstream catchments of watersheds in Dorim stream are mountainous with steep land slopes, the surface runoffs concentrate rapidly to channel flowing into the downstream urban area once the storm rainfall starts. So, flash flood frequently causes loss of human life and property in this area. Figure 5 shows the rainfall–runoff model of drainage networks in Dorim basin. In result of the constructing drainage networks (all basin of Dorim area), the number of conduits is 32,471 (Figure 5a). After simplification, the number of conduits is 243 (Figure 5b). Although the number of conduits substantially decreases as the criteria of the cumulative drainage area, it was presented that SWMM parameters was calculated automatically and basin area was the same as before the simplification.

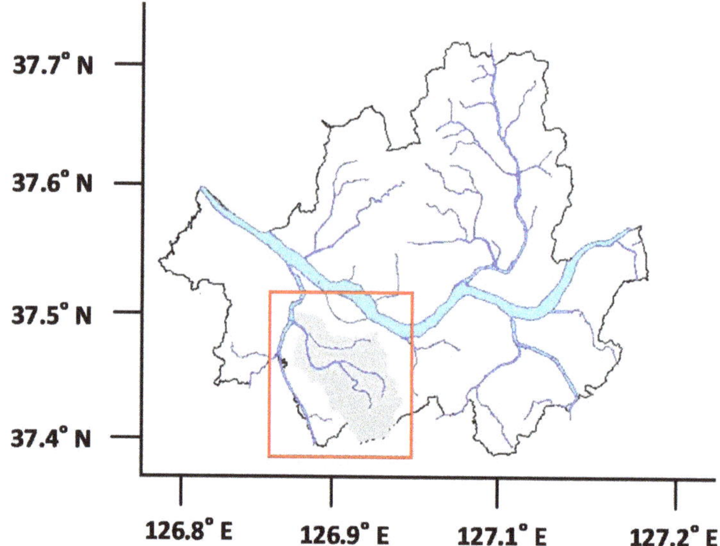

Figure 4. Map of the study area in Seoul metropolitan city.

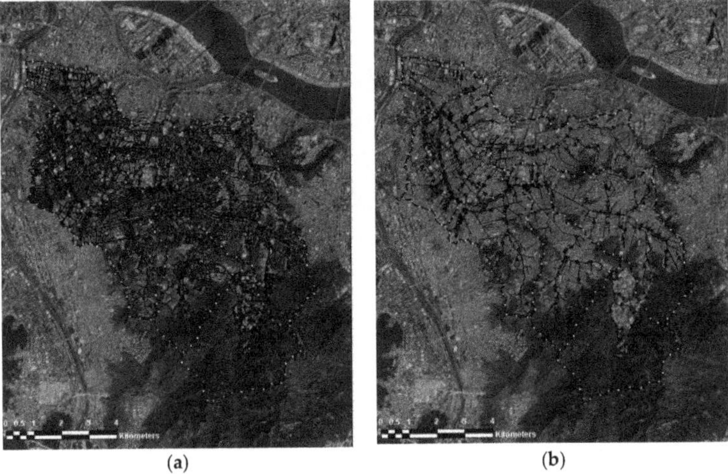

(a)   (b)

Figure 5. The simplified result of drainage networks in rainfall–runoff model of Dorim basin (**a**) before simplification; (**b**) after simplification.

3.2. Hydrological Time Series Data

For the construction of a hydrological time series data set for training LSTM, water level data in the Dorim stream were obtained from Seoul and district offices. The Seoul Metropolitan City has installed 32 water level stations at Han River and 19 monitoring systems at branch streams to record the water level. Among them, water level data from Dorim bridge, Sindaebang, and Sillim 3 bridge stations in Dorim basin are provided. Water level data from Seoul University, Gwanak Dorim bridge, Guro Digital Complex station, and Guro1 Bridge stations were also obtained from each district office. Figure 6 shows the location map of water level stations and the water level data in Dorim stream after performing quality control. Dorim stream is fed by four tributaries, Daebang stream, Bongcheon stream, and Droim 1 and 2 tributaries. Among them, Dorim 2 tributary flows into the Dorim stream located in Sillim 3 bridge, Bongcheon stream flows into Gwanak Dorim bridge, and Dorim 1 tributary flows into Guro Digital Complex station. In water level data, Dorim bridge, Sindaebang, and Guro 1 bridge was excepted from training. There were a lot of outliers in Dorim bridge station because it used to be influenced by the backwater of Ahnyang stream located downstream of Dorim stream, a lot of missing values in Sindaebang station, and a readjustment of water level station in Guro 1 bridge. Table 3 shows the statistical characteristics of water level data set.

A calibration strategy of training set, validation set, and test set size ratio is important to present statistically significant results. However, since before getting great performance, it is not easy to determine the statistical significance. Nevertheless, since approximate values of the error rates of time-series data forecasting on similar tasks are known, it is possible to estimate what reasonable size a test set should have [28]. In this study, the data sets were divided based on the number of rainfall events. Data from 2011 to 2017 were the training dataset, data in 2018 were the validation dataset, and data in 2019 were the test dataset (Figure 7).

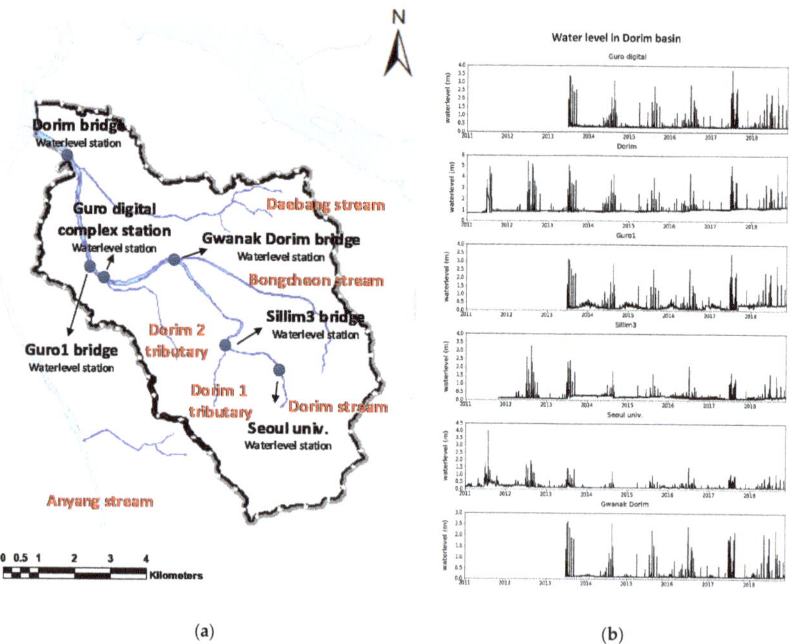

(a)   (b)

Figure 6. Locations of the Dorim basin. (**a**) Location map of water level stations in Dorim stream; (**b**) water level data with quality control in Dorim stream.

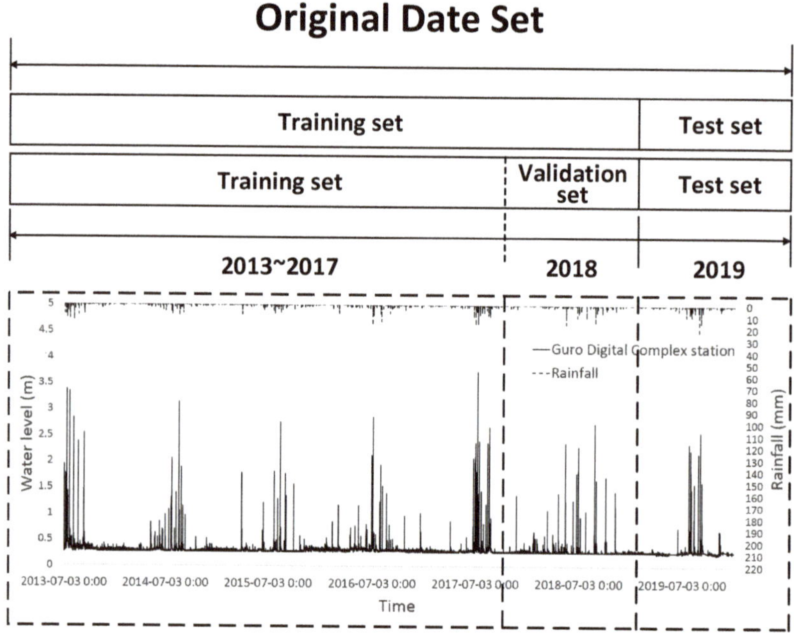

Figure 7. Building of hydrologic time series data sets (training set, validation set, test set) in the short-term inundation prediction module.

Table 3. Statistical characteristics of water level data set.

Data Set	Station	Min	Max	Avg.	STDEV
Training set	Guro Digital Complex station	0.27	3.73	0.31	0.103
	Gwanak Dorim bridge	0.10	2.62	0.12	0.069
	Sillim 3 bridge	0.12	2.42	0.21	0.087
	Seoul University	0.14	1.56	0.16	0.067
Validation set	Guro Digital Complex station	0.27	2.74	0.30	0.108
	Gwanak Dorim bridge	0.08	2.20	0.13	0.086
	Sillim 3 bridge	0.04	1.59	0.20	0.060
	Seoul University.	0.01	1.08	0.12	0.089
Test set	Guro Digital Complex station	0.23	2.57	0.29	0.091
	Gwanak Dorim bridge	0.08	1.89	0.14	0.061
	Sillim 3 bridge	0.08	1.43	0.16	0.049
	Seoul University	0.01	0.99	0.11	0.074

4. Application and Evaluation of Integrated System

4.1. Model Application

The LSTM model was trained with water level data to predict stream depth in the short-term inundation prediction module for 2019. Guro Digital Complex station was selected for the target point for predicting the stream depth and flood warning. This is the most dangerous point where accidents were frequently caused by flooding. Furthermore, rainfall forecasting using radar and numerical analysis of drainage networks in real-time were carried out in the very short-term inundation prediction module. 2-dimension inundation maps were also provided to point inundation warning area.

4.1.1. Application of Short-Term Inundation Prediction

Tributaries inflow have a substantially influence on forecasting results of water level by LSTM. Therefore, we have constructed the training sets with various input combinations to consider tributaries inflow. They showed the training group due to water level data (Table 4) and the accuracy of the forecasting result. The module of the short-term inundation prediction was carried out with LSTM on 26 July 2019. The forecasting results for Cases 1, in which training was performed using only one upstream depth data, did not reflect good forecasting model performance because other streams flow into Dorim stream. However, the forecasting results obtained with depth data of two points, case 2 showed better accuracy than case 1 result. Moreover, the forecasting results for Case 3, which involved the training with depth data of three points, indicated the best performance. It means that tributary data is very important to predict downstream depth in Dorim basin. Case 3 also presented that if sufficient depth data are acquired and the data quality is improved, the forecasting result based on depth data can reflect sufficiently good performance. Through these results, the LSTM with training three points water level stations was selected as forecasting model of short-term inundation prediction module. After quantitative and qualitative analysis of LSTM, we also scatter the observed and simulated water level values (Figure 8). The values LSTM R^2 are 0.9695, 0.9653, and 0.9823, respectively due to lead time. The LSTM has high values of R^2 indicating that this model could well reflect the relationship between observed and forecasting depth so that forecasting model with lead time 90 min is quite possible to predict and warning stream flooding. It is sufficiently can warn and evacuate the resident if accurate water level prediction with 90 min lead time is led in flash rainfall.

Table 4. Training group due to training conditions in water level data.

Case	Model	Data	Input Variable	Output Variable
Case 1			W_{se} (t), W_{se} (t − 1), …, W_{se} (t − τ)	W_{gdc} (t + 3)
Case 2	LSTM	Water level	W_{se} (t), W_{se} (t − 1), …, W_{se} (t − τ) W_{sil} (t), W_{sil} (t − 1), …, W_{sil} (t − τ)	W_{gdc} (t + 6) W_{gdc} (t + 9)
Case 3			W_{se} (t), W_{se} (t − 1), …, W_{se} (t − τ) W_{sil} (t), W_{sil} (t − 1), …, W_{sil} (t − τ) W_{gdb} (t), W_{gdb} (t − 1), …, W_{gdb} (t − τ)	

W_{gdc} = Water level data of Guro Digital Complex station, W_{gdb} = Water level data of Gwanak Dorim bridge, W_{sil} = Water level data of Sillim3 bridge, W_{se} = Water level data of Seoul University. t = Current time, τ = previous time.

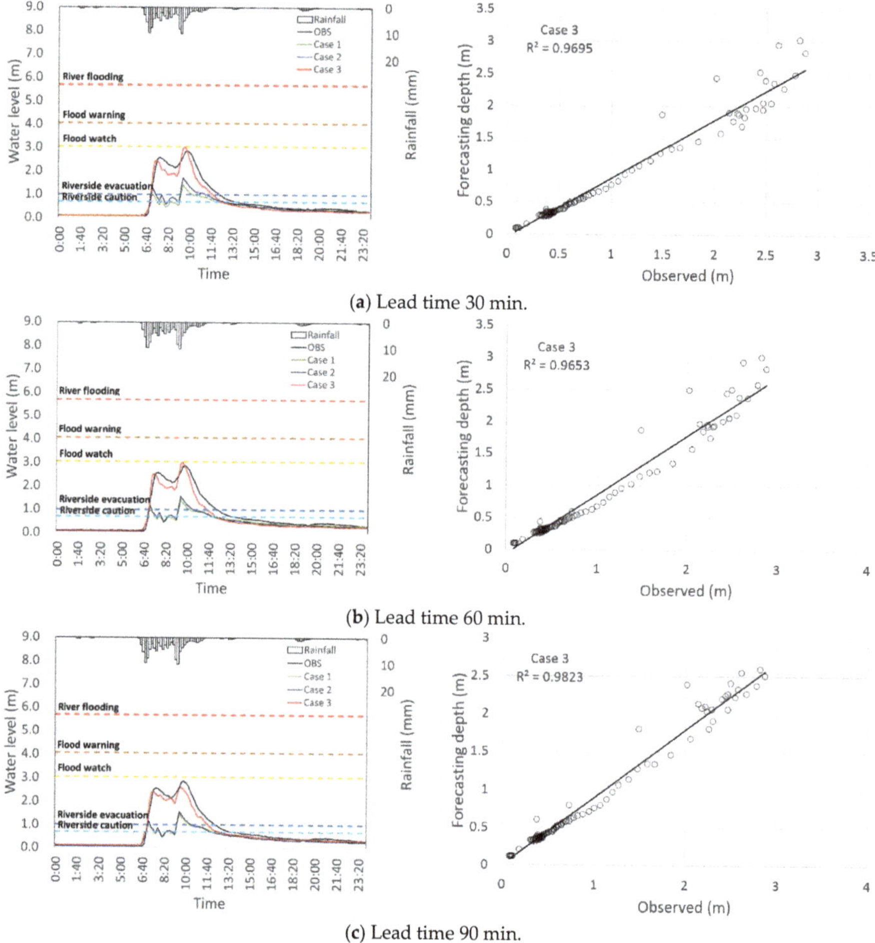

(**a**) Lead time 30 min.

(**b**) Lead time 60 min.

(**c**) Lead time 90 min.

Figure 8. Prediction results of Long Short-Term Memory (LSTM) due to varying lead times on Guro Digital Complex station and scatter plot of the observed and the simulated stream depth in training 3 water level stations. (**a**) Lead time 30 min. (**b**) Lead time 60 min. (**c**) Lead time 90 min.

4.1.2. Application of Very Short-Term Inundation Prediction

Rainfall forecasting results by radar in the very short-term inundation prediction were applied. This module generated the forecasting rainfall with from 10 min to 120 min lead time for the Automatic Weather station (AWS) 8 points of KMA near Dorim basin. After that, it converted to the areal average rainfall and the accuracy of forecasting results was compared with that of AWS rainfall data on 26 July 2019 (Figure 9). As a result, the forecasting results until 70 min were good predicted, the accuracy from 80 min began to decrease sharply. Among these forecasting results, the 40 min forecasting rainfall result predicted by radar was applied to the simplified rainfall–runoff model to calculate the stream depth and accuracy of analyzed results was examined. The analysis points were selected as the water level stations in Dorim basin (Seoul University, Sillim3 bridge, Gwanak Dorim bridge, Guro Digital Complex station, and Guro1 Bridge). Figure 10 shows that analyzed stream depth results were compared with the data of water level stations on 26 July 2019. Table 5 makes the comparison of performances of SWNN rainfall–runoff model for stream depth prediction. This is the quantitative analysis of the SWMM model using four preference criteria. Due to the characteristics of predicted rainfall results by radar, which did not predict the initial rainfall well, R^2 and NSE of the upstream where Seoul University is located were relatively low, but the predictive performance of the downstream, Gwanak Dorim, and Guro 1 bridge was 0.80 and 0.81, respectively, at 40 min lead time which presented good prediction performance. After 50 min lead time, the values of MAE and RMSE are all increased sharply. These cases illustrate that the 40 min lead time is the most appropriate time to forecast and warn urban flooding.

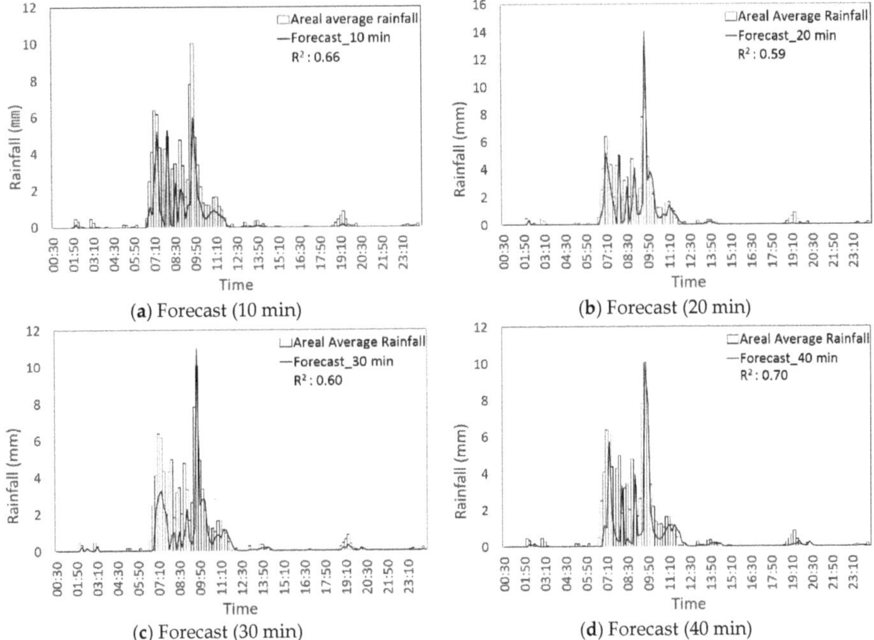

Figure 9. *Cont.*

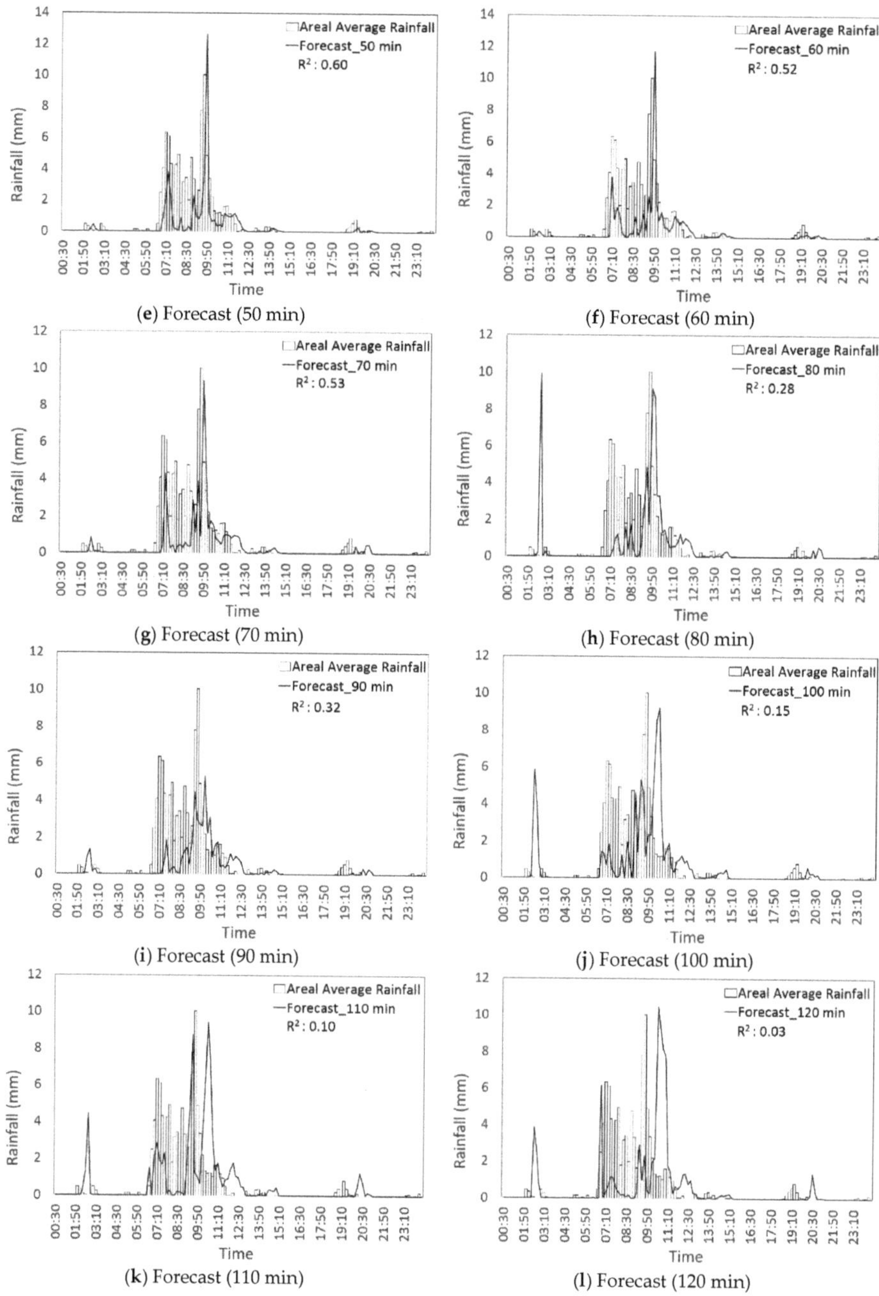

Figure 9. Predicted areal average rainfall by radar due to forecast lead time in Dorim basin. (**a**) Forecast (10 min). (**b**) Forecast (20 min). (**c**) Forecast (30 min). (**d**) Forecast (40 min). (**e**) Forecast (50 min). (**f**) Forecast (60 min). (**g**) Forecast (70 min). (**h**) Forecast (80 min). (**i**) Forecast (90 min). (**j**) Forecast (100 min). (**k**) Forecast (110 min). (**l**) Forecast (120 min).

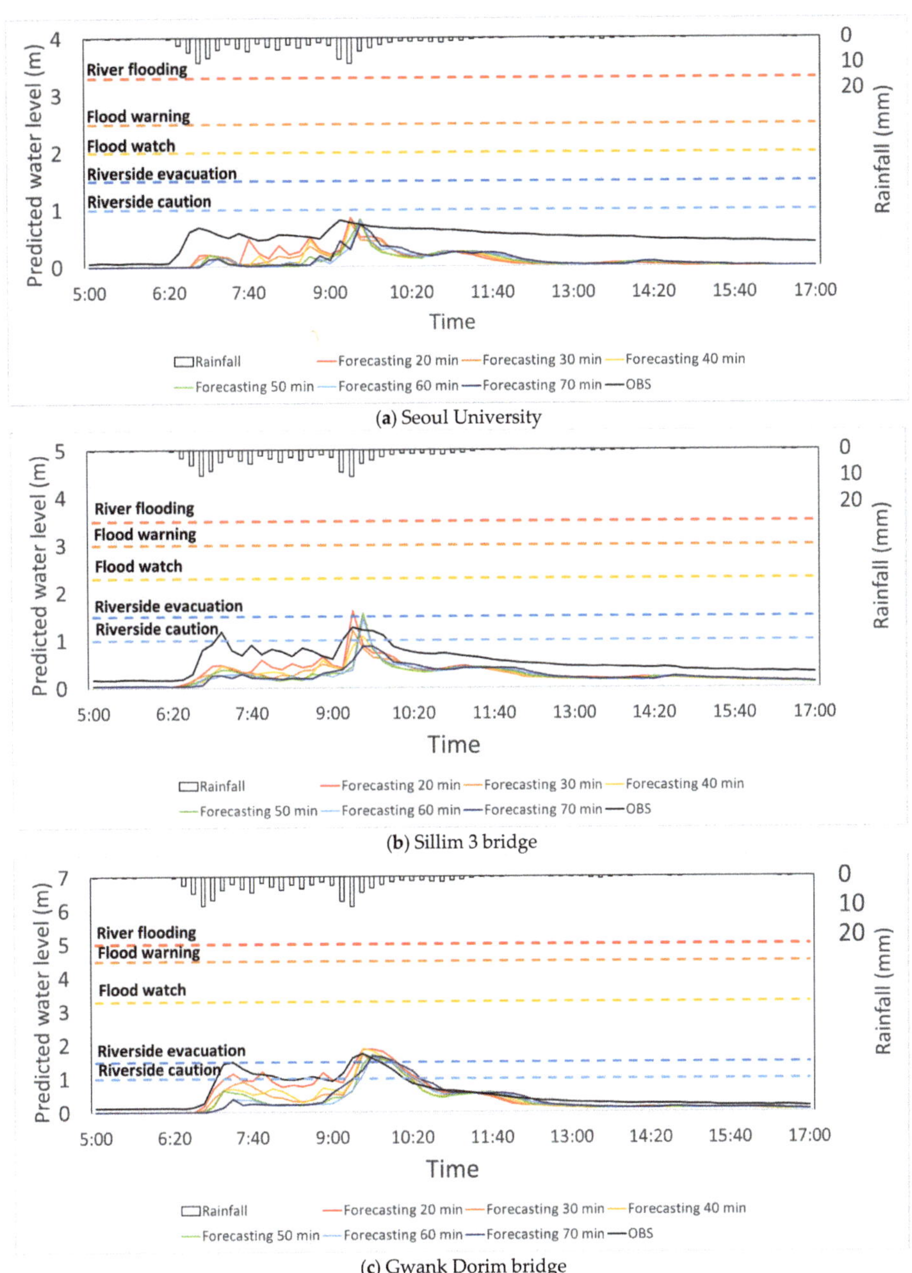

(**a**) Seoul University

(**b**) Sillim 3 bridge

(**c**) Gwank Dorim bridge

Figure 10. *Cont.*

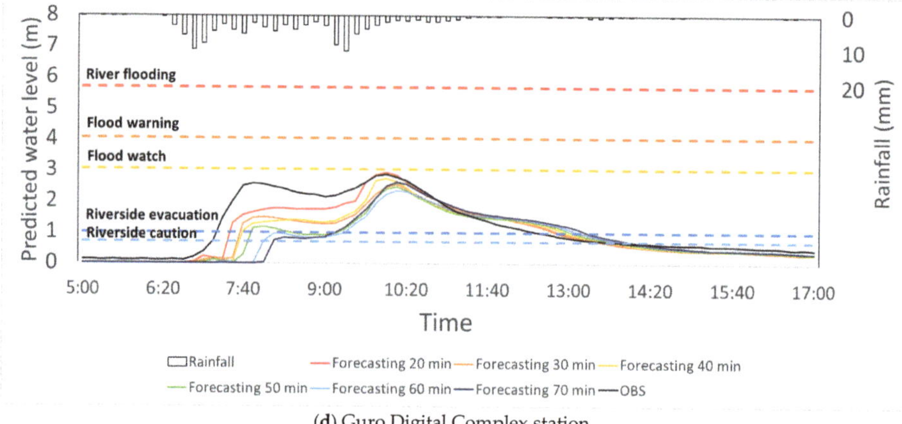

(**d**) Guro Digital Complex station

Figure 10. Rainfall–runoff results of SWMM, the predicted water level by radar forecast rainfall across Dorim stream. (**a**) Seoul University. (**b**) Sillim 3 bridge. (**c**) Gwank Dorim bridge. (**d**) Guro Digital Complex station.

Table 5. Comparison of performances of SWMM for stream depth prediction (lead time from 20 min to 70 min by radar).

Stations	R^2	Forecast 20 min			R^2	Forecast 30 min		
		RMSE	**NSE**	**MAE**		**RMSE**	**NSE**	**MAE**
Seoul University	0.35	0.34	−1.71	0.30	0.41	0.35	−1.73	0.30
Sillim 3 birdge	0.83	0.23	0.18	0.21	0.82	0.25	0.08	0.21
Gwank Dorim bridge	0.94	0.14	0.86	0.13	0.82	0.21	0.69	0.15
Guro Digital Complex station	0.93	0.40	0.48	0.35	0.85	0.46	0.31	0.38
Stations		**Forecast 40 min**				**Forecast 50 min**		
Seoul University	0.34	0.35	−1.82	0.31	0.36	0.36	−1.91	0.32
Sillim 3 birdge	0.80	0.25	0.03	0.22	0.62	0.27	−0.14	0.23
Gwank Dorim bridge	0.81	0.21	0.69	0.15	0.67	0.27	0.50	0.17
Guro Digital Complex station	0.83	0.46	0.30	0.38	0.71	0.52	0.10	0.41
Stations		**Forecast 60 min**				**Forecast 70 min**		
Seoul University	0.36	0.35	−1.89	0.31	0.37	0.35	−1.80	0.31
Sillim 3 birdge	0.61	0.27	−0.15	0.22	0.68	0.27	−0.09	0.22
Gwank Dorim bridge	0.56	0.30	0.38	0.18	0.57	0.29	0.41	0.18
Guro Digital Complex station	0.60	0.56	−0.04	0.42	0.55	0.57	−0.06	0.41

4.1.3. Evaluation of the Integrated Flood Forecasting and Warning System

The prediction of peak depth and warning time of flood is critical for flood forecasting and warning process. Therefore, to evaluate the applicability of integrated flood forecasting and warning system, the forecasting results of integrated flood forecasting and warning system were compared with River Emergency Evacuation Notification System (REENS) operated by Seoul Metropolitan Office which used real-time stream depth to warn flooding (Table 6). There was the first warning, "riverside caution" at 07:10 on 26 July 2019 in observed stream depth of Guro Digital Complex station and the second warning, "riverside evacuation" at 7:20. REENS also issued simultaneously both warnings, "riverside caution" and "riverside evacuation" at 07:10. The short-term inundation prediction modules of LSTM that have lead time from 30 min to 90 min issued exactly the riverside warning time as with observations (OBS). Moreover, early warnings were implemented with from 30 min to 90 min through training the lead time data. The very short-term inundation prediction modules of SWMM that have lead time 20 min to 40 min analyzed the riverside warning time 10-min later compared with

OBS because initial rainfall forecasting has lower performance in this rainfall event. However, early warnings were able to be implemented with from 10 min to 20 min through rainfall prediction by radar. These forecasting results illustrate that both modules can be forecasting and warning urban flooding adequately. The short-term inundation prediction module based on the data-driven method can predict the stream depth with long lead times and high-quality forecasting performance. The very short-term inundation prediction module based on the model-driven method has a slightly shorter lead time than the data-driven method because of time-consuming SWMM analysis but it presents the rainfall and inundation forecasting results across the watershed as well as 2-dimensional inundation maps for decision-making in the web pages (Figure 11) [29]. In addition, it is expected that ability of urban flood forecasting and warning in the integrated flood forecasting and warning system can be increased through improve the accuracy of rainfall prediction by radar, advancement of rainfall–runoff model and growth of observed hydrology time series data.

Table 6. Comparison of flood warning time in SWMM, LSTM, and River Emergency Evacuation Notification System (REENS).

Flood Warning	Riverside Caution	Riverside Vacuation
Observation	7:10	7:20
REENS	7:10	7:10
LSTM (30 min lead time)	6:50 (7:20)	6:50 (7:20)
LSTM (60 min lead time)	6:20 (7:20)	6:20 (7:20)
LSTM (90 min lead time)	5:40 (7:10)	5:50 (7:20)
SWMM (20 min lead time)	7:00 (7:20)	7:00 (7:20)
SWMM (30 min lead time)	7:00 (7:30)	7:00 (7:30)
SWMM (40 min lead time)	6:50 (7:30)	6:50 (7:30)

( ) = Warning time by forecasting analysis.

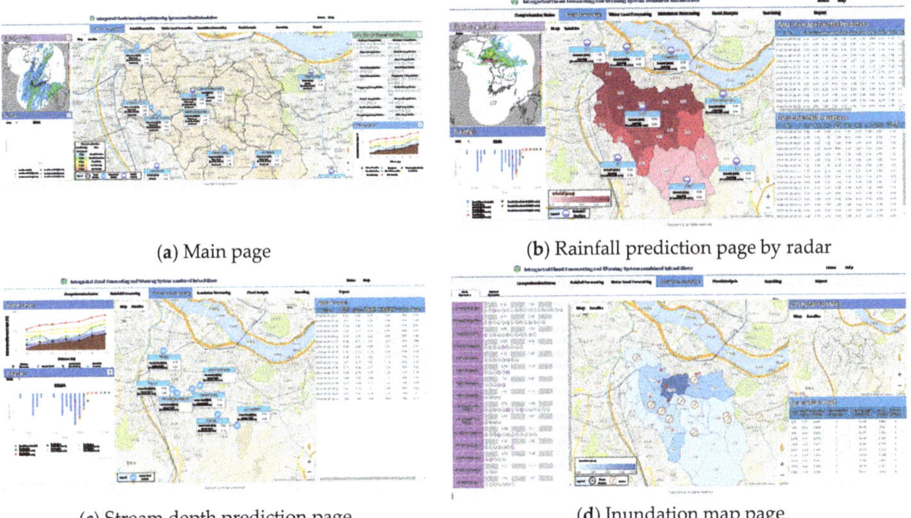

(a) Main page

(b) Rainfall prediction page by radar

(c) Stream depth prediction page

(d) Inundation map page

Figure 11. Integrated flood forecasting and warning system in web page Ref [29]. (**a**) Main page. (**b**) Rainfall prediction page by radar. (**c**) Stream depth prediction page. (**d**) Inundation map page.

5. Conclusions

A real-time flood forecasting and warning can proactively convey a powerful message about the warning of flash floods and inundation depths in urban areas, together with the effectiveness of possible countermeasures to citizens and civil servants. Therefore, we developed the integrated flood forecasting and warning system combined inland-river that systematized technology to quantify flood risk and flood forecasting to provide effective flood warning in the urban area. It consisted of the short-term inundation prediction module with data-driven method and the very short-term inundation prediction module with model-driven method. LSTM was used to predict the stream depth in the short-term inundation prediction and very short-term rainfall prediction by radar data, construction of a rainfall–runoff model combined inland-river by coupled SWMM and HEC-RAS, automatic simplification module of drainage networks, automatic calibration module of SWMM parameter by DDS, construction of the 2-dimension inundation database by various rainfall scenario to warn inundation risk areas were used in very short-term inundation prediction to warn and convey the flood-related data and information to communities.

An application to Dorim basin which is one of the Seoul Metropolitan's inundation vulnerable areas is performed to clearly demonstrate the superiority of the proposed system. First, according to the LSTM model in the short-term inundation prediction module, it is observed that the stream depth in the study area is predicted well from 30 min to 90 min lead time. Second, in the very short-term inundation prediction module, rainfall from 10 min to 70 min lead time was predicted with a good performance by radar. Moreover, these results are provided to the rainfall–runoff model of SWMM with an automatic simplification method and DDS optimization algorithm applied. Forecasting results of stream depth across the river are also predicted with great performance until 40-min lead time. Finally, 2-dimension inundation maps are presented through the inundation database which is made due to various rainfall and duration scenarios. The proposed model can be improved through improving the accuracy of rainfall prediction by radar, the advancement of the rainfall–runoff model, and the growth of observed hydrology time series data. We expect that it can be effectively applied to reduce human and property damage caused by flood.

Author Contributions: Conceptualization, resources, formal analysis, and writing—original draft, J.H.L.; conceptualization, methodology, and writing—review and editing, Y.-I.M.; and writing—review and editing, G.M.Y. and H.T.M. All authors have read and agreed to the published version of the manuscript.

Funding: This work was supported by the 2018 Research Fund of the University of Seoul.

Acknowledgments: The University of Seoul has financially supported this research. The authors would like to thank the anonymous reviewers for their valuable comments and suggestions to improve the quality of the paper.

Conflicts of Interest: The authors declare no conflict of interest.

References

1. Karamouz, M.; Hosseinpour, A.; Nazif, S. Improvement of urban drainage system performance under climate change impact: Case study. *J. Hydrol.* **2010**, *16*, 395–412. [CrossRef]
2. Wu, X.S.; Wang, Z.L.; Guo, S.L.; Liao, W.L.; Zeng, Z.Y.; Chen, X.H. Scenario-based projections of future urban inundation within a coupled hydrodynamic model framework: A case study in Dongguan City, China. *J. Hydrol.* **2017**, *547*, 428–442. [CrossRef]
3. IPCC. *Climate Change 2014: Synthesis Report*; Pachauri, R.K., Meyer, L.A., Eds.; IPCC: Geneva, Switzerland, 2014; pp. 1–151.
4. Kim, S.M.; Tachikawa, Y.; Takara, K. Recent flood disasters and progress of disaster management system in Korea. *Annu. Disaster Prevent. Res. Inst. Kyoto Univ.* **2007**, *50*, 15–31.
5. Blong, R. A Review of Damage Intensity Scales. *Nat. Hazards* **2003**, *29*, 57–76. [CrossRef]
6. Seoul Metropolitan City. River Emergency Evacuation Notification System. 2013. Available online: http://203.142.217.120/ (accessed on 19 August 2020).
7. Lin, G.F.; Lin, Y.H.; Chou, Y.C. Development of a real-time regional-inundation forecasting model for the inundation warning system. *J. Hydroinform.* **2013**, *15*, 1391–1407. [CrossRef]

8. Chang, L.C.; Chang, F.J.; Yang, S.N.; Kao, I.F.; Ku, Y.Y.; Kuo, C.L.; Amin, I.M.Z.M. Building an intelligent hydroinformatics integration platform for regional flood inundation warning systems. *Water* **2018**, *11*, 9. [CrossRef]

9. Berne, A.; Delrieu, G.; Creutin, J.D.; Obled, C. Temporal and spatial resolution of rainfall measurements required for urban hydrology. *J. Hydrol.* **2004**, *299*, 166–179. [CrossRef]

10. Bruni, G.; Reinoso, R.; Giesen, N.C.; Clemens, F.H.L.R.; Veldhuis, J.A.E. On the sensitivity of urban hydrodynamic modelling to rainfall spatial and temporal resolution. *Hydrol. Earth Syst.* **2015**, *19*, 691–709. [CrossRef]

11. Einfalt, T.; Arnbjerg-Nielsen, K.; Golz, C.; Jensen, N.E.; Quirmbachd, M.; Vaes, G.; Vieux, B. Towards a roadmap for use of radar rainfall data in urban drainage. *J. Hydrol.* **2004**, *299*, 186–202. [CrossRef]

12. Foresti, L.; Reyniers, M.; Seed, A.; Delobbe, L. Development and verification of a real-time stochastic precipitation nowcasting system for urban hydrology in Belgium. *Hydrol. Earth Syst.* **2016**, *12*, 6831–6879. [CrossRef]

13. Shi, X.; Gao, Z.; Lausen, L.; Wang, H.; Yeung, D.Y.; Wong, W.K.; Woo, W.C. Deep Learning for Precipitation Nowcasting: A Benchmark and a New Model. Available online: http://papers.nips.cc/paper/7145-deep-learning-for-precipitation-nowcasting-a-benchmark-and-a-new-model (accessed on 19 August 2020).

14. Chang, F.J.; Chen, P.A.; Lu, Y.R.; Huang, E.; Chang, K.Y. Real-time multi-step-ahead water level forecasting by recurrent neural networks for urban flood control. *J. Hydrol.* **2014**, *517*, 836–846. [CrossRef]

15. Kim, H.U.; Bae, T.S. Preliminary study of deep learning-based precipitation prediction. *J. Korean Soc. Surv.* **2017**, *35*, 423–430.

16. Hu, C.; Wu, Q.; Li, H.; Jian, S.; Li, N.; Lou, Z. Deep learning with a long short-term memory networks approach for rainfall-runoff simulation. *Water* **2018**, *10*, 1543. [CrossRef]

17. Kang, J.; Wang, H.; Yuan, F.; Wang, Z.; Huang, J.; Qiu, T. Prediction of precipitation based on recurrent neural networks in Jingdezhen. *Atmosphere* **2020**, *11*, 246. [CrossRef]

18. Song, T.; Ding, W.; Wu, J.; Liu, H.; Zhou, H.; Chu, J. Flash flood forecasting based on long short-term memory networks. *Water* **2019**, *12*, 109. [CrossRef]

19. Bergstra, J.; Bengio, Y. Random search for hyper-parameter optimization. *J. Mach. Learn. Res.* **2012**, *13*, 281–305.

20. Hochreiter, S.; Schmidhuber, J. Long short-term memory. *Neural Comput.* **1997**, *9*, 1735–1780. [CrossRef]

21. Kingma, D.; Ba, J. Adam: A Method for Stochastic Optimization. Available online: https://arxiv.org/abs/1412.6980 (accessed on 19 August 2020).

22. He, K.; Zhang, X.; Ren, S.; Sun, J. Delving Deep into Rectifiers: Surpassing Human-Level Performance on Imagenet Classification. Available online: https://www.cv-foundation.org/openaccess/content_iccv_2015/html/He_Delving_Deep_into_ICCV_2015_paper.html (accessed on 19 August 2020).

23. Atlas, D.; Ulbrich, C.; Marks, F.D.; Amitai, E.; Williams, C.R. Systematic variation of drop size and radar-rainfall relations. *J. Geophys. Res.* **1999**, *104*, 6155–6169. [CrossRef]

24. Rinehart, R.E.; Garvey, T. Three dimensional storm motion detection by conventional weather radar. *Nature* **1978**, *273*, 287–289. [CrossRef]

25. Kim, G.; Kim, J.P. Development of a short term rainfall forecast model using sequential CAPPI data. *J. Korean Soc. Civ. Eng.* **2009**, *29*, 543–550. (In Korean)

26. Tolson, B.A.; Shoemaker, C.A. Dynamically dimensioned search algorithm for computationally efficient watershed model calibration. *Water Resour. Res.* **2007**, *43*, W01413. [CrossRef]

27. Huff, F.A. Time distribution of rainfall in heavy storm. *Water Resour. Res.* **1967**, *3*, 1007–1019. [CrossRef]

28. Guyon, I.; Makhoul, J.; Schwartz, R.; Vapnik, V. What size test set gives good error rate estimates? *IEEE Trans. Pattern Anal. Mach. Intell.* **1998**, *20*, 52–64. [CrossRef]

29. University of Seoul, Integrated Flood Forecasting and Warning System Combined Inland-River. 2018. Available online: http://210.125.181.27:10001/IRFF/index.-do (accessed on 19 August 2020).

MDPI
St. Alban-Anlage 66
4052 Basel
Switzerland
Tel. +41 61 683 77 34
Fax +41 61 302 89 18
www.mdpi.com

Atmosphere Editorial Office
E-mail: atmosphere@mdpi.com
www.mdpi.com/journal/atmosphere